공업영어

김지환 · 김 중 · 김도석 · 이창훈 공저

북스힐

머리말

글로벌시대를 사는 현대인들에게 외국어는 꼭 필요하다. 그리고 최근 과학기술이 급속하게 발전하면서 과학기술에 관한 지식이나 정보를 영어로 이해하고 표현할 필요성이 많아졌다. 그래서 이 책은 그런 정보를 필요로 하는 공과대학이나 전문대학 학생을 위하여 엮은 것이다.

과학적 기술적 사실을 표현하는 언어라고 해도 일상생활에서 쓰는 언어와 다른 것은 아니며, 단지 어휘의 사용상의 문제로 그 뜻 자체가 대체로 특수화되고 전공분야에 따라서 전문화된 다른 어떤 뜻을 갖고 사용될 뿐이다.

일반적으로 고등학교까지의 과정에서 기본이 되는 문법을 배웠으므로, 이 책의 내용은 문법을 다루기보다는 공학적인 내용의 문장을 통하여 전문용어를 습득하는 데 목적을 두었다.

이 책에 나오는 전문용어들은 KS에 규정된 용어에 따랐다.

이 책은 19장으로 구성되어 있다. 문장의 내용들은 일반적으로 공학에서 다루는 과목들 중 기초적인 내용이 대부분이지만 뒷부분의 일부 내용들은 예전에 없던 내용들을 넣어서 꾸몄다. 특히 micro/nano technology, mechatronics, robotics 등의 내용을 소개하였다.

계획은 오래전부터 했으나 집필을 시작하고 보니 의욕만이 앞서고 내용이 그에 따르지 못한 점을 깊이 느꼈다.

자고나면 바뀌는 세상에 새로 나오는 과학적인 내용들을 모두 다루지는 못하겠지만 학생들이 공학을 공부하는데 필요한 내용이나 용어들을 앞으로 계속 보완하고자 한다.

끝으로 이 책을 만드는데 많은 도움을 주신 윤여권교수와 북스힐 여러분들에게 깊은 감사를 드린다.

2008. 8

저 자

차 례

01 Solid Mechanics

Hooke's law in figure. 1-1, first stage OA is called the "elastic" stage. The second stage BC usually has some undulation in the diagram as shown, but it is substantially a horizontal piece of curve at constant force, and the elongation will increase from 0.001 in/in at B to about 0.020 in/in at C. This second stage is known as the "plastic" stage, in which the stress is substantially constant and equal to the yield stress.

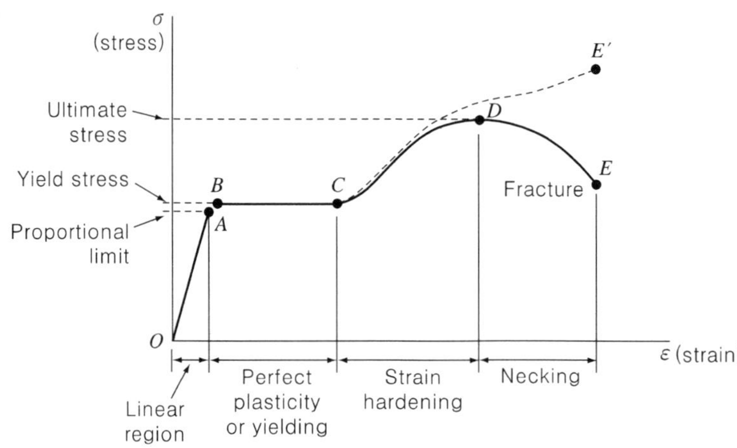

[Figure 1-1] Stress-strain diagram

Note

- Hooke's law : 훅크의 법칙
- elastic stage : 탄성영역
- substantially : 사실상, 대체로
- elongation : 신장, 연장
- plastic stage : 소성 영역
- yield stress : 항복응력
- first stage : 제1단계
- undulation : 파동, 진동 음파
- horizontal piece : 수평 부분
- increase : 증가하다(시키다)
- constant : 변치않는, 일정한
- stress : 응력, 스트레스

- strain : 변형률, 뒤틀게하다
- diagram : 선도, 그림
- proportional limit : 비례한도
- fracture : 파괴
- strain hardening : 변형경화
- ultimate stress : 극한 응력

cf. elastic ↔ plastic
elasticity : 탄성 ↔ plasticity : 소성

In the third stage CE, the deformations become very large, and test piece "necks", or shows a locally diminished diameter, and finally breaks in the center of necking at point D in the diagram. If the load is let off at some point between B and E, before the final failure, the return curve is substantially a straight line parallel to OA, as shown by dotted lines, and in that case the bar shows a "permanent elongation" after release of the load to zero. Our subject of strength of materials deals with only with the material in the first, or elastic stage.

Note

- deformation : 변형
- test piece : 시험편
- neck : 목, 잘록한 부분, 해협
- locally diminished diameter : 국부적으로 감소된 지름
- center of necking : 지름이 감소된 부분의 중심
- load : 부하, 하중
- load test : 하중시험
- failure : (재료의)파손, 파단
- final failure : 최후 파단
- return curve : 회복 곡선
- dotted line : 점선
- permanent elongation : 영구 변형
- after release of the load : 하중 제거 후
- subject : 목적
- strength : 강도, 세기
- deal with : ~를 다루다. ~를 취급하다
- material : 물질, 재료
- strength of materials : 재료역학

The elastic, linear behavior of a material in the region OA of figure. 1-1 is known as "Hooke's law". It states that the elongation is proportional to the force, or expressed in a formula

$$\sigma = \frac{P}{A} \qquad E = \frac{\sigma}{\varepsilon}$$

The quantity $\varepsilon = \triangle l / l$ is the "unit elongation" or "strain" expressed in meters per meter and hence dimensionless. The quantity σ is the stress or the force per unit area, expressed in Newtons per square meter. Finally, E is a proportionality constant, which also must be expressed in Newtons per square meter. It is known as the "modulus of elasticity", or also as "Young's-modulus" .

Note

- behavior(-viour) : (재료의) 거동
- region : 영역, 범위
- proportional to : ~에 비례하여
- express : 표현하다, 나타내다, (속달로) 보내다
- formula : 식, 공식
- quantity : 양(量) *cf.* quality : 품질, 성질
- unit elongation : 단위 신장량
- stress : 응력
- dimensionless : 무차원의
- unit area : 단위 면적
- proportionality : 비례
- constant : 상수
- modulus of elasticity : 비례상수

  cf. Newton : 힘의 단위(과학자 Newton의 이름)

  ε ; 변형률, E ; 탄성계수, young 률

Shear stress

A stress, of force per unit cross-sectional area, is called a "normal stress" when the direction of the force is perpendicular to the plane of the cross section. Such normal stresses are either tensile or compressive, depending on their sign. When the stress lies in the plane of the cross section, it is called a "shear stress".

Note

- shear stress : 전단응력
- cross-sectional : 단면의
- normal stress : 수직응력
- perpendicular to ~ : ~에 수직인

- tensile : 인장의
- compressive : 압축의
- sign : 기호

  cf. tension : 인장 ↔ compression : 압축

  normal stress : tensile stress(+), compressive stress(-)

Simple construction elements in which shear stress occur are shown in figure. 1-2 the first of these shows two boiler plates connected by a riveted lap joint. If the rivet is not in tension and hence the two plates do not press against each other, there is no friction between them and a tensile force between the plates must be transmitted as a shear stress across the mid-section of the shank of the rivet.

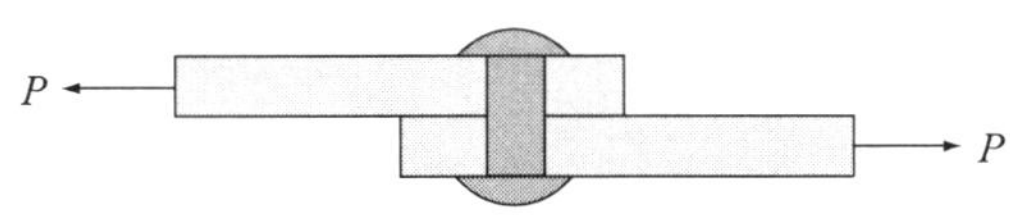

[Figure 1-2] Shear stress

Note

- construction element : 구조 요소
- lap joint : 겹치기 이음
- friction : 마찰
- be transmitted : 전달되다
- mid-section : 단면의 중앙
- shank : 몸통, 자루

Poisson's ratio

When a sample of material is stretched in one direction, it tends to contract (or rarely, expand) in the other direction. Poisson's ratio (ν), is a measure of this tendency. Poisson's ratio is the ratio of the relative contraction strain, or transverse strain (normal to the applied load), divided by the relative extension strain, or axial strain (in the direction of the applied load).

Assuming that the material is compressed along the axial direction:

$$\nu = -\frac{\varepsilon_x}{\varepsilon_y}$$

ν is resulting Poisson's ratio

ε_x is transverse strain(negative for axial tension, positive for axial compression)

ε_y is axial strain(positive for axial tension, negative for axial compression).

Note

- Poisson's ratio : 프와송의 비
- be stretched : 늘어나다
- contract : 수축하다
- ratio : 비, 비율
- transverse : 가로축의, 교축의
- axial direction : 축방향
- positive : 양의, 적극적인
- direction : 방향
- tend to ~ : ~하는 경향이 있다
- measure : 척도, 측정
- relative : 상대적인
- strain : 변형, 변형량
- negative : 마이너스, 음, 부정적인

Fatigue

Many structures and machine element in daily use are subjected to alternating stress, so that the fact that such parts will eventually break under an alternating load two to four times smaller than the steady load of ultimate failure is of very great practical importance. Under alternating stress, the material is said to fatigue and fatigue-strength properties are determined experimentally by the simple fatigue test machine.

Note

- fatigue : 재료의 피로
- structure : 구조, 구조물, 건축물

- machine element : 기계요소
- alternating load : 변동하중
- steady load : 정상하중
- fatigue test machine : 피로 시험기
- property : 설질 특성
  cf. of + 명사 = 명사의 형용사형의 뜻
  of importance= important.
- alternating stress : 변동응력
- two to four times : 2~4배
- ultimate failure : 극한 파손
- experimental : 실험의, 실험에 의해

Torsion and bending

In engineering mechanics, bending(also known as flexure) characterizes the behavior of a structural element subjected to an external load applied perpendicular to the axis of the element. A structural element subjected to bending is known as a beam. A closet rod sagging under the weight of clothes on clothes hangers is an example of a beam experiencing bending.

A torsion is the twisting of an object due to an applied torque. In circular sections, the resultant shearing stress is perpendicular to the radius.

Note

- torsion : 비틀림
- engineering mechanics : 공업역학
- characterize : 특징을 나타내다
- axis : 축, 축선
- closet : 벽장
- sag : 휘다
- resultant : 결과적인
- shearing stress : 전단응력
- radius : 반지름, 편심거리(radii : 복수)
- bending : 굽힘
- flexure : 휨
- perpendicular : 수직
- beam : 보
- rod : 봉
- twisting : 뒤틀림
- subjected to ~ : ~를 받는

Stress Concentration

In a structure or machine part having a notch or any abrupt change in cross section, the maximum stress will occur at this location and will be greater than the stress calculated by elementary formulas based upon simplified assumptions as to the stress distribution. The ratio of this maximum stress to the nominal stress (calculated by the elementary formulas) is the stress-concentration factor K_t.

This is a constant for the particular shape and is independent of the material, provided it is isotropic. The stress-concentration factor may be determined experimentally or, in many cases, theoretically from the mathematical theory of elasticity.

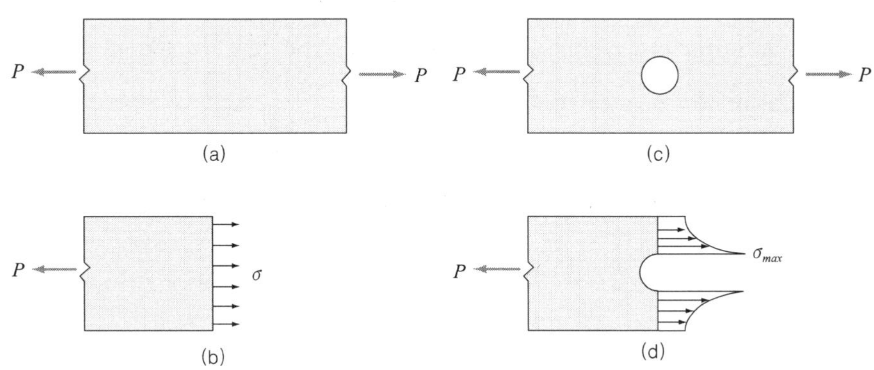

[Figure 1-3] Stress Concentration

Note

- stress concentration : 응력 집중
- notch : V자 모양의 홈 또는 틈
- abrupt : 갑작스로운, 뜻밖의
- cross section : 단면
- maximum stress : 최대응력
- nominal stress : 공칭응력
- ratio : 비, 비율
- stress-concentration factor : 응력집중계수
- theory of elasticity : 탄성률

Thermal stress

Thermal stress and strain are caused by temperature change. Stresses due to the restriction of thermal expansion or contraction of a body are called thermal stresses. In a statically determinate structure, a uniform temperature change will not cause any stresses, as deformations are permitted to occur freely. On the other hand, a temperature change in a structure that is supported in a statically intermediate manner will induce stresses in the member.

The equation of thermal strain and stress are as follows.

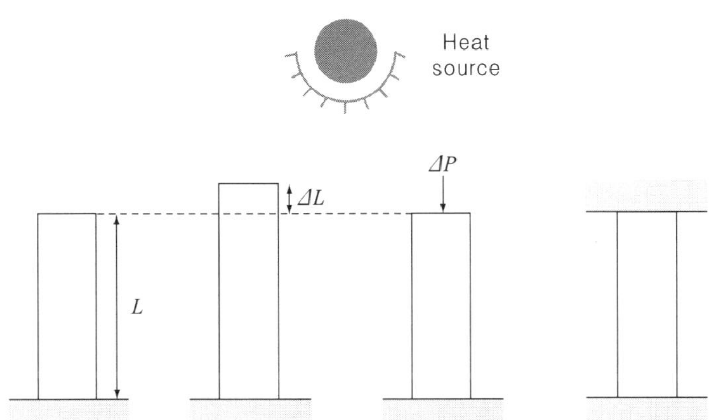

[Figure 1-4] Thermal stress

$$\Delta L = \alpha \cdot \Delta t \cdot L$$

$$\varepsilon = \alpha \cdot \Delta t$$

$$\sigma_{th} = E \cdot \alpha \cdot \Delta t$$

Where

σ_{th} is thermal stress

α is thermal coefficient

Δt is temperature increase(+) / decrease(−)

L is initial length

Note

- Thermal stress : 열응력
- expand : 늘어나다
- restricting : 제한된
- restrained : 구속된
- temperature change : 온도변화
- statically determinate structure : 정정구조물
- statically indeterminate : 부정정
- thermal coefficient : 열팽창계수

02 Engineering Materials

Metals

The most widely used metallic elements are iron, copper, lead, zinc, aluminum, tin, nickel, and magnesium. Some of these are used extensively in the pure state, but by far the largest amount is consumed in the form of alloys. Metals may be divided according to their characteristics as the heavy metals(iron, copper, nickel, etc.), light metals(aluminum and magnesium), white metal(lead, tin, zinc, cadmium, etc.), and precious metals(gold, platinum, palladium, silver, etc.)

Note

- metallic element : 금속원소
- iron : 철
- copper : 구리
- lead : 납
- zinc : 아연
- aluminum : 알루미늄
- tin : 주석
- magnesium : 마그네슘
- pure state : 순수한 상태
- alloy : 합금
- heavy metal : 중금속 ; 비중 5.0 이상
- light metal : 경금속, 비중 4이하
- white metal : 백색합금
- precious metal : 귀금속
- gold : 금(원소기호 Au)
- platinum : 백금(원소기호 Pt)
- palladium : 팔라디움(원소기호 Pd)
- silver : 은(원소기호 Ag)

Irons

The earth contains a large number of metals which are useful to man. One of the most important of these is iron.

The iron ore which we find in the earth is not pure. It contains some impurities which we must remove by smelting. The process of smelting consists of heating the ore in a blast furnace with coke and limestone, and reducing it to metal. Blasts or hot air enter the furnace from the bottom and provide the oxygen which is necessary for the reduction of the ore. The ore becomes molten, and its oxides combine with carbon form the coke. The non-metallic constituents of the ore combine with the limestone to form a liquid slag. The floats on top of the molten iron, and passes out of the furnace through a tap the metal which remain is pig-iron.

Note

- iron ore : 철광석
- impurity : 불순물
- smelting : 제련
- consist of ~ : ~로 구성되다
- blast furnace : 용광로(鎔鑛爐), 고로(高爐)
- coke : 코크스
- blast : 블라스트, 송풍(送風)
- oxygen : 산소
- oxide : 산화물
- constituent : 성분, 요소
- carbon : 탄소(원소기호 C)
- limestone : 석회석, 석회암
- slag : 슬래그(광석의 용재)
- pig-iron : 선철(銑鐵)

Cast iron

We can melt this down again in another furnace a cupola with more coke and limestone, and tap it out into a ladle or directly into mould. This is cast iron. Cast iron does not have the strength of steel. It is brittle and may fracture under tension. But it possesses certain properties which make it useful in the manufacture of machinery. In the molten state it is very fluid, and therefore it is easy to cast it into intricate shape. Also it is ease to machine it. Cast iron contains small proportions of other substances. These non-metallic constituents of cast iron include carbon, silicon and sulphur, and the presence of these substances affects the behavior of the metal.

Note

- cast iron : 주철
- limestone : 석회석
- mould : 형틀, 주형(鑄型)
- tension : 인장력
- intricate shape : 복잡한 형상
- cupola : 용선로(鎔銑爐), 큐폴라
- ladle : 레이들, 주조용 쇳물바가지
- brittle : 부서지기 쉬운
- molten state : 융해된 상태
- sulphur : 유황 (=sulfer)

The carbon in cast iron is present partly as free graphite and partly as a chemical combination of iron and carbon which we call cementite. This is a very hard substance, and it makes the iron hard too. However, iron can only hold about 1.5% of cementite. Any carbon content above that percentage is present in the form of flaky graphite. Steel contains no free graphite, and its carbon content ranges from almost nothing to 1.5%. We make wire and tubing from mild steel with a very low carbon content, and drills and cutting tools from high carbon steel.

Note

- graphite : 흑연
- flaky : 박편(薄片)의
- content : 함유량
- cutting tool : 절삭공구
- cimentite : 시멘타이트, 탄화철
- steel : 강철, 강
- mild steel : 연강, 저탄소강
- high carbon steel : 고탄소강

Nonferrous metals

Aluminum is a silvery-white metal with a melting point of 659. 7℃. It is low in weight and has a high strength-weight ratio. It has a specific gravity of 2.70.

Copper is a pink-colored metal with a melting point of 1083℃. For its excellent conductivity of electricity and heat, its resistance to corrosion, and its ductility and strength, copper is used very widely. Near its melting point, it becomes extremely brittle. Under the influence of bismuth, lead, oxygen or sulphur it becomes brittle and under the influence of arsenic, antimony, iron, nickel or oxygen its conductivity deteriorates. It has a specific gravity of 8.94

Nickel is a silvery-white metal with a specific gravity of 8.90. it melts at a temperature of 1,455℃. Nickel is used as an alloying element in steel and nonferrous metals, and produces a marked increase in the toughness and corrosion resistance of the alloyed metal. It is also widely used as an electroplated coating for steel and other metals.

Lead is a exceptionally heavy, soft and malleable metal. It is of blue-gray color. It has a specific gravity of 11.34 and melts at 327.4℃. As it is of low tenacity of a ton per square inch it cannot be drawn.

Zinc is a white-colored metal with a gray tinge. It has a specific gravity of 7.14 and melts at 419.5℃. The principal applications of zinc are in the galvanizing of steel for corrosion resistance and the manufacture of die casting. It is also used in making brass, paints, batteries and other products.

Tin is a white meal with a yellowish tint, melts at 231.9℃, and has specific gravity of 7.30. It is very soft and malleable. It can be rolled into very thin sheets. Tin is a primary ingredient of solder, bronze, babbit, collapsible tube, etc.

Note

- non-ferrous metal : 비철금속
- melting point : 녹는점, 용해점
- specific gravity : 비중
- copper : 구리

- conductivity : 전도성
- electricity : 전류
- resistance : 저항
- bismuth : 비스무트, 창연(원소기호 Bi)
- brittle : 깨지기 쉬운
- oxygen : 산소
- sulphur : 유황
- arsenic : 비소(원소기호 As)
- arsenic : 비소
- antimony : 안티몬(원소기호 Sb)
- corrosion : 부식
- deteriorate : 저하되다
- toughness : 인성
- malleable : 가단성(可鍛性)의
- galvanizing : 갤버나이징, 아연피복
- ingredient : (혼합물의) 성분
- babbit : 배빗
- collapsible : 접을 수 있는

Alloys

Stainless steels contain alloying elements such as chromium, nickel, etc. which promote the ability of the metal to resist corrosion. Resistance to the action of strong acids and ability to remain bright in highly corrosive environments are characteristics of stainless steel. Some stainless steels are extremely hard, some have unusual ductility, and some are valuable because they retain their strength at very high temperatures.

The most principal alloying element in promoting rust resistance is chromium.

Other elements used to impart special properties to the steel are nickel, molybdenum, titanium, columbium, zirconium, etc.

Note

- alloy : 합금
- chromium : 크롬(원소기호 Cr)
- nickel : 니켈, 백통(원소기호 Ni)
- promote : 촉진시키다
- corrosion : 부식, 녹
- acid : 산, 산성
- environment : 환경
- ductility : (재료의) 연성, 잡아늘이기 쉬운

Bronzes and brasses are well-known alloys of copper.

Tin additions to copper form the basis of bronze and zinc additions form the basis of brass.

Additions of aluminum, phosphorus, silicon, manganese, etc., with other metals yield special alloy, such as the aluminum-bronze, phosphor-bronzes, silicon-bronzes and manganese-bronzes.

Copper, zinc and nickel in proper combination produce the well-known nickel-silver alloy, and 70 % copper with 30% nickel produces cupro-nickel.

Note

- bronze : 청동
- brass : 놋쇠, 황동
- phosphorus : 인(원소기호 P)
- cupro-nickel : 백통

The best known of aluminum alloys is of the **duralumin** type. This has a composition of 4.5% copper, 0.6% manganese and 1.5% magnesium. Duralumin has the advantage of being very light. It is approximately one-third the weight of steel, and yet it can be made almost as strong as steel by heat treatment.

White metal is a series of alloys based on tin, lead or antimony. Because or their antifriction properties, these alloys are used primarily in bearings. Another use of these alloy is in the production of printing type. Babbit metal is one of these alloys. The term "white metal" is applied also to many other alloys that have whitish color.

Note

- duralumin : 듀랄루민
- one-third : $\frac{1}{3}$
- heat treatment : 열처리
- white metal : 백색합금
- antimony : 안티몬(원소기호 Sb)
- babbit metal : 배비트 메탈(주석 70~90%에 안티몬 등을 가한 베어링 합금)

Medium carbon steel

Steels containing from 0.3 to 0.83% carbon are probably the most widely used steels for the construction of equipment. Those steels which contained less than approximately 0.35% carbon did not harden appreciably by quenching. the properties of those steels containing 0.35% carbon or more are greatly improved by heat-treatment and, therefore, are best adapted for machine parts. For these purposes, it is customary to use steels containing between 0.35 and 0.55% carbon. Increased carbon content lowers the ductility of steel. Therefore, when greater toughness is required, a lower-carbon may be necessary, and the desired strength can be obtained through suitable heat-treatment

Note

- medium carbon steel : 중탄소강
- appreciably : 다소, 얼마간
- quenching : 담금질

  cf. heat treatment
  - quenching : 퀜칭, 담금질
  - annealing : 어닐링, 풀림
  - tempering : 템퍼링, 뜨임
  - normalizing : 노멀라이징, 불림
- toughness : 강도, 단단함
- strength : strong, 강도
- ductility : (재료)의 연성 잡아늘리기 쉬운 성질
- brittle : 취성, 변형이 적어 파괴되기 쉬운 성질

Molybdenum steel

The resultant effects of molybdenum in steel may be summarized as follows

(1) Higher tempering temperatures are required to develop mechanical properties equivalent to the plain carbon steel.

(2) Greater ductility and toughness may be obtained for given elastic limit.

(3) Greater machinability at higher hardness is obtained.

(4) Greater hardenability is obtained, particularly when chromium is present.

(5) Temper brittleness is greatly reduced or eliminated.

(6) Creep resistance at high temperatures is improved because of the stability of the carbide

Note

- molybdenum steel : 몰리브덴강
- be summarized : 요약되다
- tempering temperature : 뜨임온도
- plain carbon steel : 보통탄소강
- machinability : 기계가공성
- hardenability : 경화성
- temper brittleness : 뜨임 취성
- creep : 크리프(일정한 온도에서 일정한 하중이 작용할 때, 시간이 경과함에 따라 재료의 변형이 증가하는 현상
- stability : 안정성
- carbide : 탄화물

Ductility

Ductility is a mechanical property which describes how able the material lends itself to be formed into rod-like shapes before fracture occurs. Examples of highly ductile metals are silver, gold, copper, and aluminium. The ductility of steel varies depending on the alloying constituents. Increasing levels of carbon decreases ductility, i.e. the steel becomes more brittle.

Brittleness

A material is brittle if it is liable to fracture when subjected to stress. That is, it has little tendency to deform (or strain) before fracture. This fracture absorbs relatively little energy, even in materials of high strength, and usually makes a snapping sound.

Hardness is the characteristic of a solid material expressing its resistance to permanent deformation.

Note

- lend itself to ~ : ~의 구실을 하다, ~에 적합하다
- ductile metal : 연성 금속
- alloying constituent : 합금성분
- be liable to ~ : ~하기 쉬운
- fracture : 파괴
- permanent deformation : 영구 변형

  cf. ductility(연성) ↔ brittleness(취성)

Copper

Copper is also the base of a very important class of alloys. A large amount of copper is used for tubing in water lines, for fuel and oil lines in aircraft, and for conducting food stuffs, beverages, pulpmill pulp, chemicals, etc. in view of excellent thermal conductivity, it is used for kettles, etc. Copper is also very resistant to atmospheric corrosion and is, therefore, used for roofing, window screens, etc.

Note

- tubing : 배관
- fuel : 연료
- oil line : 유압배관
- foodstuff : 식료품
- beverage : 음료
- thermal conductivity : 열전도율
- atmospheric corrosion : 대기 부식

Season cracking of brass

Brasses that contain more than 20% zinc are susceptible to spontaneous cracking and to corrosion at grain boundaries in the cold-worked condition. After severe cold-work, as in deep drawing operation, high internal stresses are established in the material.

Crack may develop under this condition after a time; hence, the term is season cracking. Season cracking can be prevented in these alloys by a low temperature anneal which tend to decrease the residual or internal stresses in the materials.

Note

- season cracking : 시즌균열, 자연(自然)균열
- be susceptible to ~ : ~에 민감하다
- corrosion : 부식, 녹
- grain boundary : 입계
- cold-work : 금속을 냉각 가공하다
- deep drawing : 디프드로잉하다(판금가공시 다이스에 밀어넣고 컵이나 상자모양을 가공하다
- internal stress : 내부응력
- crack : 균열
- residual stress : 잔류응력

Leaded brass

The machinability of most of the non-ferrous metals and alloys can be improved by the addition of lead. The addition of up to about 4% lead improves the machinability of brass quite appreciably. Such an addition tends to reduce the ductility.

Note

- leaded brass : 납을 함유한 황동
- non-ferrous : 비철금속
- alloy : 합금

Titanium

This metal has entered the commercial light-metal field through extensive research on processing method to decrease the cost of production. The cost, however, is still high for common use or the materials. Improved methods of manufacture together with new applications permitting large-scale production will tend to lower the price as has been the case with both aluminum and magnesium. Several desirable properties place titanium between aluminum and the stainless steels as a light, strong, ductile and corrosion-resistant metal. Titanium is unique in combining a high melting point of 1726 ℃ with a low specific gravity of 4.5. It has a corrosion resistance equivalent to 18% chromium-8% nickel stainless steel.

Note

- commercial : 상업용의
- light metal field : 경금속계
- high melting point : 고융용점
- unique : 유일한
- specific gravity : 비중

White cast iron

White cast iron is not employed for structural parts because of its excessive brittleness. However, it is useful in resisting wear by abrasion. The dissociation of iron carbide of gray cast iron can be prevented by rapid cooling, hence, the use of a chill in the mould will form white cast iron at the surface, thus producing a gray cast iron at the surface, thus producing a gray cast iron with a hard, wear-resistant surface. The handness and brittleness increase as carbon content increases.

Note

- White cast iron : 백철광
- brittleness : 취성
- abrasion : 마모, 벗겨짐
- gray cast iron : 회주철
- chill : 주조물의 표면이 주조 때 급냉되어 경백선철로 되는 것
  cf. chill mold : 칠주형, 주물표면을 급냉하여 경도를 높이는데 사용하는 주형

Ceramic

The term covers inorganic non metallic materials which are formed by the action of heat. Up until the 1950s or so, the most important of these were the traditional clays, made into pottery, bricks, tiles and the like, along with cements and glass. Clay-based ceramics are described in the article on pottery. A composite material of ceramic and metal is known as cermet. The word ceramic can be an adjective, and can also be used as a noun to refer to a ceramic material, or a product of ceramic manufacture. Ceramics may also be used as a singular noun referring to the art of making things out of ceramic materials.

Many ceramic materials are hard, porous, and brittle. The study and development of ceramics includes methods to mitigate problems associated with these characteristics, and to accentuate the strengths of the materials as well as to investigate novel applications.

Note

- Ceramic : 세라믹, 도기
- inorganic : 무기물의
- clay : 점토
- pottery : 도기, 오지그릇
- cermet : 서멧, 내열성 합금
- singular noun : 단수 명사
- composite material : 복합재료
- porous : 작은 구멍이 많은
- mitigate : 완화시키다, 가볍게 하다
- accentuate : 두드러지게 하다

Composite materials

Composite materials(or composites for short) are engineering materials made from two or more constituent materials with significantly different physical or chemical properties and which remain separate and distinct on a macroscopic level within the finished structure.

Composite materials have gained popularity(despite their generally high cost) in high-performance products that need to be lightweight, yet strong enough to take harsh loading conditions such as aerospace components(tails, wings, fuselages, propellers), boat and scull hulls, bicycle frames and racing car bodies. Other uses include fishing rods and storage tanks. The new Boeing 787 structure including the wings and fuselage is composed of over 50 percent composites.

Note

- composite material : 복합 재료
- engineering material : 공업 재료
- macroscopic : 거시적인
- popularity : 대중성
- harsh loading condition : 극한 하중 상태
- aerospace : 우주공간, 항공우주산업
- scull hull : (조정용) 보트 선체
- bicycle frame : 자전거 구조
- fishing rod : 낚싯대
- fuselage : 비행기 동체

Semiconductor

A semiconductor is a solid material that has electrical conductivity in between that of a conductor and that of an insulator; it can vary over that wide range either permanently or dynamically. Semiconductors are tremendously important in technology. Semiconductor devices, electronic components made of semiconductor materials, are essential in modern electrical devices. Examples range from computers to cellular phones to digital audio players. Silicon is used to create most semiconductors commercially, but dozens of other materials are used as well.

Note

- semiconductor : 반도체
- electrical conductivity : 전기 전도성
- conductor : 도체(導體)
- insulator : 부도체(不導體)

- tremendously : 엄청나게, 매우
- cellular phone : 핸드폰
- dozens of ~ : 수십의, 수많은

Indentation hardness

Primarily used in engineering and metallurgy, indentation hardness seeks to characterize a material's hardness; i.e. its resistance to permanent, and in particular plastic deformation. It is usually measured by loading an indenter of specified geometry onto the material and measuring the dimensions of the resulting indentation.

There are several alternative definitions of indentation hardness, the most common of which are

(1) Brinell hardness test (HB);
(2) Janka Wood Hardness Rating;
(3) Knoop hardness test (HK) or micro-hardness test, for measurement over small areas;
(4) Meyer hardness test;
(5) Rockwell hardness test (HR)
(6) Shore durometer hardness, used for polymers;
(7) Vickers hardness test (HV), has one of the widest scales;
(8) Barcol hardness test, for composite materials, scale from 0 to 100.

The Vickers and Brinell hardness scales correlate well over a wide range, however, with Brinell only producing overestimated values at high loads.

Hardness increases with decreasing particle size. This is known as the Hall-Petch effect.

Note

- indentation : 오목, 압흔, 경도시험시 강구에 의해 시험면이 남겨지는 오목한 흔적
- indentation hardness : 압입 경도
- metallurgy : 야금학
- indenter : 압입자
- correlate : 서로 관계있다

Creep

Creep is the term used to describe the tendency of a solid material to slowly move or deform permanently under the influence of stresses. It occurs as a result of long term exposure to levels of stress that are below the yield strength or ultimate strength of the material. Creep is more severe in materials that are subjected to heat for long periods, and near the melting point. It is often observed in glasses. Creep is a monotonically increasing function of temperature.

The rate of this deformation is a function of the material properties, exposure time, exposure temperature and the applied load (stress). Depending on the magnitude of the applied stress and its duration, the deformation may become so large that a component can no longer perform its function. Creep is usually of concern to engineers and metallurgists when evaluating components that operate under high stresses or high temperatures.

Note

- creep : 일정한 온도에서 일정한 하중이 작용할 때 시간이 경과함에 따라 재료의 변형이 증가하는 현상.
- stress : 응력
- yield strength : 항복강도
- ultimate strength : 극한강도
- melting point : 녹는점, 융해점
- monotonically : 단순하게
- material property : 재료의 성질

03 Fluid Mechanics

Density

In physics the density (ρ) of a body is the ratio of its mass (m) to its volume (V), a measure of how tightly the matter within it is packed together. Its SI units are kilograms per cubic meter (kg/m^3). It is also sometimes given in the cgs units of grams per cubic centimeter (g/cm^3).

Density is defined by:

$$\rho = \frac{M}{V}$$

Note

- physics : 물리학
- density : 밀도
- mass : 질량
- volume : 체적
- SI unit : International system of units, 국제단위계
- cubic meter : 입방미터, 체적의 단위(m^3)
- cgs : centimeter-gram-second를 기본단위로 하는 미터법 절대단위계

Specific gravity

Specific gravity is defined as the ratio of the density of a given substance to the density of water, when both are at the same temperature, it is therefore a dimensionless quantity. Substances with a specific gravity greater than one are denser than water, and so (ignoring surface tension effects) will sink in it, and those with a specific gravity of less than one are less dense than water, and so will float in it. Specific gravity is a special case of, or in some usages synonymous with, relative density, with the latter term often preferred in modern scientific writing. The use of specific gravity is discouraged in technical use in scientific fields requiring high precision−actual density (in dimensions of mass per unit volume) is preferred.

Specific gravity, SG, is expressed mathematically as:

$$SG = \frac{\rho_{substance}}{\rho_{H_2O}}$$

where ρ is the density of the substance, and is the density of water. (By convention ρ, the Greek letter rho, denotes density.) The density of water varies with temperature and pressure, and it is usual to refer specific gravity to the density at 4°C and a normal pressure of 1 atm. In this case ρ is equal to 1000 kg/m^3 in SI.

Note

- specific gravity : 비중
- substance : 물질
- dimensionless quantity : 무차원량
- surface tension : 표면장력
- sink : 가라앉다.
- float : 뜨다.
- synonymous : 같은 뜻의
- relative density : 비밀도
- atm : 표준대기압을 나타내는 단위, 1 atm = 101,325 Pa

Specific weight

The specific weight (also known as the unit weight) is the weight per unit volume of a material, or:

$$\gamma = \rho g$$

where γ is the specific weight of the material(weight per unit volume, typically N/m^3 units), ρ is the density of the material (mass per unit volume, typically kg/m^3), g is acceleration due to gravity (rate of change of velocity, given in m/s^2) .

Specific weight can be used in mechanical engineering to determine the weight of a structure designed to carry certain loads while remaining intact and remaining within limits regarding deformation. It is also used in fluid dynamics as a property of the fluid.

Note

- specific weight : 비중량
- density : 밀도
- acceleration due to gravity : 중력으로 인한 가속도, 중력가속도
- mechanical engineering : 기계공학
- intact : 손상되지 않은
- deformation : 변형
- fluid dynamics : 유체역학

Static pressure

If rectangular vessel be filled with a liquid at rest, total downward force exerted by the liquid on the horizontal base of the vessel will be equal to the total weight of the liquid. Let this weight = W. If A be the area or the base, then the force per unit area exerted on the base by the liquid will be $P = W/A$. This force per unit area, P is known as the "unit pressure" or the static pressure.

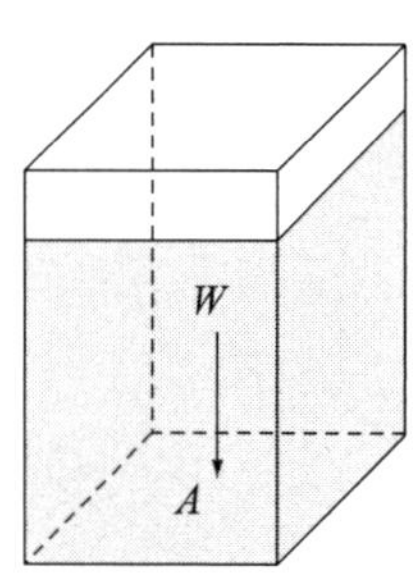

[Figure 3-1] Static pressure

Note

- static pressure : 정압
- rectangular vessel : 직사각형 용기
- downward : 아랫방향의
- unit area : 단위면적
- intensity of pressure : 압력의 강도, 세기

Static head

The vertical distance h between any selected point in a liquid at rest and the free surface exposed to the atmosphere is spoken of as the head of liquid above the point. Alternatively we might say that at the selected point the head or the static head or the pressure head is h the connection between head and intensity of pressure is given by following equation.

$$h = \frac{p}{\gamma}$$

Note

- static head : 정수두
- vertical distance : 수직거리
- at rest : 정지 상태의
- exposed to ~ : ~에 노출된

Pressure measurement

The absolute pressure reaches zero when an ideal vacuum is achieved, that is , when no molecules are left in space; consequently, a negative absolute pressure is impossibility. A second scale is defined by measuring pressures relative to the local atmospheric pressure. This pressure is so-called gage pressure. A conversion from gage pressure to absolute can be carried out using the following equation;

$$absolute\ pressure = gagepressure + atmosphere\ pressure$$

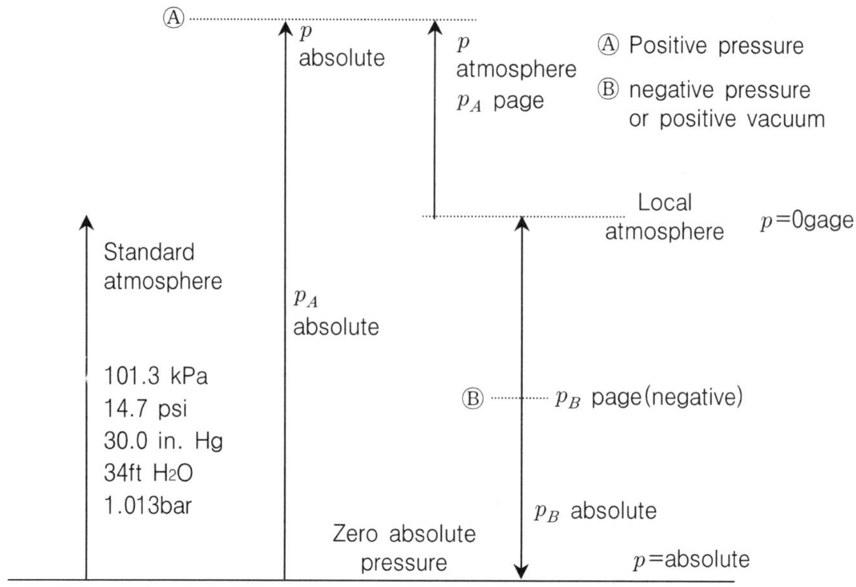

[Figure 3-2] gage pressure and absolute pressure

Note

- absolute pressure : 절대압력
- atmospheric pressure : 대기압
- gage pressure : 계기압력
- consequently : 결과적으로
- so-called : 소위, 이른바

Pressure transmission

If a force F_1 acts via an area A_1 on a closed liquid, a pressure p is produced which extends throughout the whole of the liquid(Pascal's law). The same pressure applies at every point of the closed system(see figure. 3-3).

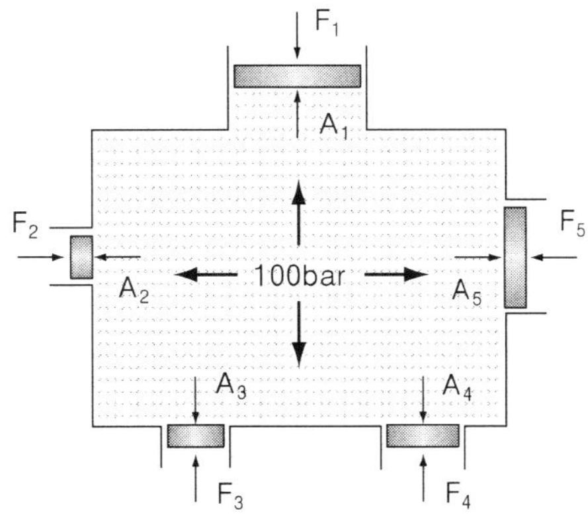

[Figure 3-3] Power transmission

Note

- closed liquid : 밀폐된 액체

Bulk modulus

Although for many purposes liquids may be regarded incompressible, yet they do in fact yield slightly under pressure. The characteristic of the liquid which determines the amount of elastic yield is termed bulk modulus of the liquid; it is comparable with the modulus of elasticity of solid and is denoted by symbol K.

$$\frac{dV}{V} = \frac{dP}{K}$$

Where,

V is original volume of liquid

dP is change of pressure

dV is change of volume corresponding to change of pressure

Note

- bulk modulus : 체적탄성률
- incompressible : 비압축성의
- elastic yield : 탄성항복
- modulus of elasticity : 탄성계수
- solid : 고체

Buoyancy

The law of buoyancy, known as Archimedes' principle, dates back some 2200 years to the Greek philosopher Archimedes. Legend has it that Hiero, king of Syracuse, suspected that his new gold crown may have been constructed of materials other than pure gold. So he asked Archimedes to test it. Archimedes probably made a lump of pure gold that weighted that same as the crown. The lump was discovered to weigh more in water than the crown weighed in water, thereby proving to Archimedes that the crown was not pure gold. The fake material possessed a larger volume to have the same weight as gold, hence it displaced more water.

Archimedes' principle is : there is buoyancy force on an object equal to the weight of displaced liquid.

Note

- buoyancy : 부력
- legend : 설화
- suspect : 의심하다
- pure gold : 순금
- lump : 덩어리
- fake : 위조하다
- possess : 가지다

Laminar flow

A viscous flow can be classified as either a laminar flow or a turbulent flow. In a laminar flow, the fluid flows with no significant mixing of neighboring fluid particles. If dye were injected into the flow, it would retain its identity for a relatively long period of time. Viscous shear stresses always influence a laminar flow. The flow may be highly time dependent.

Note

- laminar flow : 층류유동
- turbulent flow : 난류유동
- time dependent : 시간에 종속인
- viscous flow : 점성유동
- dye : 염료

Turbulent flow

This term is used in two senses. In discussing problems of pipe flow, it is convenient to describe the motion of the liquid as turbulent if it departs in the slightest degree from complete regularity or steadiness. In this sense unsteady flow is always classed as turbulent flow. There is also more general sense.

iquid motion is said to be turbulent when the elements are in extreme state of disarray, and when the velocity fluctuations are particularly violent and erratic; rotary or eddying movement are nearly always strongly developed.

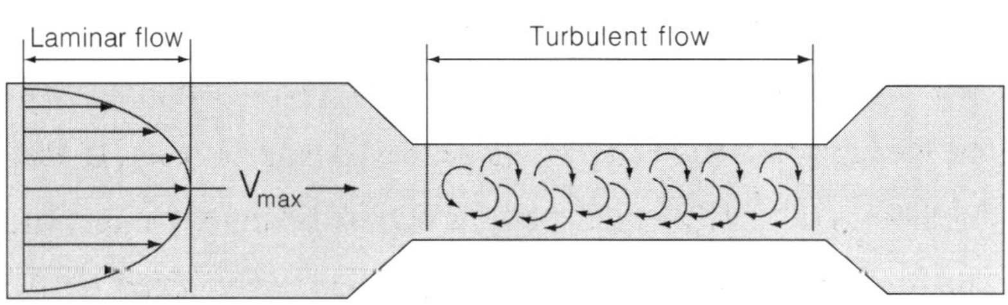

[Figure 3-4] Laminar and turbulent flow

Note

- term : 용어, 단어
- complete regularity : 완전한 규칙성
- disarray : 무질서
- erratic : 엉뚱한, 부정한
- in two senses : 두 가지 의미로
- steadiness : 정상성
- velocity fluctuation : 속도 변동
- eddying : 소용돌이

Reynolds number

The flow regime depends on three physical parameters describing the flow condition. The first parameter is length scale of the flow field, such as the thickness of a boundary layer or the diameter of a pipe. If this length scale is sufficiently, a flow disturbance may increase and the flow may be turbulent. The second parameter is a velocity scale such as a spatial average of the velocity; for a large enough velocity the flow may be turbulent. The third parameter is the kinematic viscosity; for a small enough viscosity the flow may be turbulent.

The three parameters can be combined into a single parameter that can serve as a tool to predict the flow regime. This quantity is the Reynolds number, a dimensionless parameter, defined as

$$\mathrm{Re}=\frac{vL}{\nu}$$

Where L and v are a characteristic length and velocity, respectively, and ν is the kinematic viscosity: for example, in a pipe flow L could be the pipe diameter and v could be the average velocity. If Reynolds number is relatively small, the flow regime is laminar: if it is large, the flow regime is turbulent. This is more precisely stated by defining critical Reynolds number, $\mathrm{Re}_{\mathrm{crit}}$ so that the flow is laminar if $\mathrm{Re} < \mathrm{Re}_{\mathrm{crit}}$. For example, in a flow inside a rough-walled pipe it is found that $Re_{crit} = 2000$.

Note

- Reynolds number : 레이놀즈 수
- flow regime : 유동 상황
- depend on : 기인하다, 의존하다
- boundary layer : 경계층
- spatial : 공간의
- kinematic viscosity : 동점성
- turbulent : 난류
- dimensionless parameter : 무차원 변수

- laminar : 층류
- critical Reynolds number : 임계 레이놀즈수

Conversion of energy

One of the fundamental principles of fluid mechanics, "Bernoullis Theorem" states that in such a system the total energy of a perfect liquid under ideal conditions does not change as it flows from point to point; that is, that any selected point in the system, the sum of position energy, pressure energy, and velocity energy will be equal to the sum at any other point. The equation is as follows

$$\frac{v_1^2}{2g}+\frac{p_1}{\gamma}+h_1=\frac{v_2^2}{2g}+\frac{p_2}{\gamma}+h_2$$

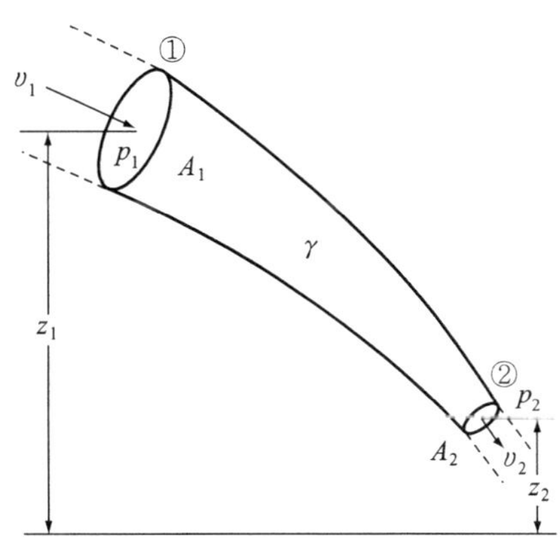

[Figure 3-5] Conservation law of flow

Note

- conversion of energy : 에너지 보존
- fundamental principle : 기본적인 원리
- ideal : 이상적인
- position energy : 위치에너지
- pressure energy : 압력에너지
- velocity energy : 속도에너지

04 Thermodynamics

Definition of thermodynamics

Thermodynamics is one of the broadest subdivisions of physical science. Strictly speaking, thermodynamics is involved with all energy transfers to or from matter. The theory and applications of thermodynamics are confined to that part of the subject necessary to a study of machines which convert heat into work or transfer heat from one substance to another.

Note

- thermodynamics : 열역학
- broadest subdivision : 가장 광범위한 분야
- physical science : 물리과학
- theory : 이론
- energy transfer : 에너지 전달
- matter : 물질
- definition : 정의
- substance : = matter, 물질

cf. convert A into B : A를 B로 바꾸다.

Specific heat

Specific heat capacity, also known simply as specific heat, is the measure of the heat energy required to increase the temperature of a unit quantity of a substance by a certain temperature interval.

The symbols for specific heat capacity are either C or c depending on how the quantity of a substance is measured. In the measurement of physical properties, the term "specific" means the measure is a bulk property (an intensive property), wherein the quantity of substance must be specified.

Note

- specific heat : 비열
- unit quantity : 단위량
- symbol : 기호

Internal energy

In thermodynamics, the internal energy of a thermodynamic system, or a body with well-defined boundaries, denoted by U, or sometimes E, is the total of the kinetic energy due to the motion of molecules (translational, rotational, vibrational) and the potential energy associated with the vibrational and electric energy of atoms within molecules or crystals. It includes the energy in all the chemical bonds, and the energy of the free, conduction electrons in metals.

The internal energy is a thermodynamic potential and for a closed thermodynamic system held at constant entropy, it will be minimized.

Note

- internal energy : 내부에너지
- boundary : 경계
- vibrational : 진동의
- crystal : 결정
- chemical bond : 화학적 결합
- closed thermodynamic system : 밀폐 열역학계
- entropy : 엔트로피

Entropy

The concept of entropy was developed in the 1850s by German physicist Rudolf Clausius who described it as the transformation-content, i.e. dissipative energy use, of a thermodynamic system or working body of chemical species during a change of state. In contrast, the first law of thermodynamics, formalized through the heat-friction experiments of James Joule in 1843, deals with the concept of energy, which is conserved in all processes; the first law, however, lacks in its ability to quantify the effects of friction and dissipation. Entropy change has often been defined as a change to a more disordered state at a molecular level.

Note

- dissipative : 소산(消散)하는
- change of state : 상태변화
- in contrast : 대조적으로
- disordered state : 무질서 상태

Enthalpy

In thermodynamics, one of the most frequently recurring is the sum of internal energy and the pressure-volume product. Since internal energy is customarily measured in cal. and the product of pressure and volume is in Newton-meters, conversion factor A = (1/427) cal/N · m is required. The sum is given the name enthalpy and is defined by

$$H = U + APV$$

Note

- enthalpy : 엔탈피
- frequently : 빈번하게
- recurring : 되풀이되어 나오는 것
- sum : 합, 총계
- pressure-volume product : 압력, 체적 곱

Freezing point, boiling point and latent point

When a solid melts or a liquid freezes, the temperature remains constant and is always the same for given substance at a given pressure. The temperature is called the freezing point

The definite quantity of heat involved in the change is called latent heat of fusion and is dependent upon pressure of temperature. When liquid boils or a vapor condenses the same characteristics appear. The temperature is called boiling point, and the quantity of heat involved, the latent heat of vaporization. These quantities vary widely with different substance.

Note

- freezing point : 어는점
- boiling point : 끓는점
- latent heat : 잠열
- melt : 녹다, 용해되다
- fusion : 융해, 용해
- vapor : 증기
- characteristic : 특징
- vaporization : 기화, 증발작용

Zeroth law of thermodynamics

When two systems are put in contact with each other, there will be a net exchange of energy between them unless or until they are in thermal equilibrium, that is they contain the same amount of thermal energy for a given volume (say, 1 cubic centimeter, or 1 cubic inch.). While this is a fundamental concept of thermodynamics, the need to state it explicitly as a law was not perceived until the first third of the 20th century, long after the first three laws were already widely in use, hence the zero numbering. The Zeroth Law asserts that thermal equilibrium, viewed as a binary relation, is an equivalence relation.

Note

- zeroth law of thermodynamics : 열역학 제 0법칙
- in contact with ~ : ~와 접촉해 있는
- thermal equilibrium : 열적 평형
- cubic inch : 세제곱인치, in^3
- assert : 주장하다
- equivalence relation : 등가관계

1st law of thermodynamics

The first law is essentially the law of conservation of energy and may be stated;

Energy can neither be created nor destroyed, only altered in form. It must be emphasized that this statement has not been proved. The result of many experiments in energy transfer and conservation are in complete agreement. However, infinite number of such experiments would still not constitute proof. In fact, a certain group of radioactive transformation seems to result in destruction.

Note

- first law of thermodynamics : 열역학 제1법칙
- essentially : 본질적으로
- conservation : 보존
- be emphasized : 강조되다
- statement : 설명
- experiment : 실험
- infinite : 무한한
- proof : 증명
- radioactive : 방사성의
- transformation : 변환, 변태
- destruction : 파괴

cf. It must be emphasized that this statement has not been proved.
실제 주어는 that 이하 절이지만, 주어부가 길 경우 가짜 주어 it을 쓰고 진짜 주어를 뒤에 배치한다.

Boyle's law

Boyle's law (sometimes referred to as the Boyle-Mariotte law) is one of the gas laws and basis of derivation for the ideal gas law, which describes the relationship between the product pressure and volume within a closed system as constant when temperature and moles remain at a fixed measure; both entities remain inversely proportional.

Note

- derivation : 도출, 유도, 미분
- ideal gas law : 이상기체법칙
- relationship : 관계

Charles's law

In thermodynamics and physical chemistry, Charles's law is a gas law and specific instance of the ideal gas law, which states that:

At constant pressure, the volume of a given mass of an ideal gas increases or decreases by the same factor as its temperature (in Kelvin) increases or decreases.

Note

- physical chemistry : 물리화학
- maintain : 유지하다

Ideal gas law

An ideal gas is defined as one in which all collisions between atoms or molecules are perfectly elastic and in which there are no intermolecular attractive forces. One can visualize it as a collection of perfectly hard spheres which collide but which otherwise do not interact with each other. In such a gas, all the internal energy is in the form of kinetic energy and any change in internal energy is accompanied by a change in temperature.

Note

- be defined as ~ : ~로써 정의되다
- atom : 원자
- molecule : 분자
- intermolecular attractive force : 분자 사이의 인력
- visualize : 시각화하다
- hard sphere : 강구(鋼球)
- collide : 충돌하다

Avogadro's law

The law stated by Avogadro is;

Equal volumes of all gases at the same pressure and temperature contain the same number of molecular. This, like the perfect-gas law, is an approximation which is applicable under the same conditions. It is useful since it permits a direct comparison between different gases.

Note

- perfect-gas law : 완전가스법칙
- approximation : 근사, (~ 에) 가까움
- molecular weight : 분자량

2nd law of thermodynamics

The second law of thermodynamics is an expression of the universal law of increasing entropy, stating that the entropy of an isolated system which is not in equilibrium will tend to increase over time, approaching a maximum value at equilibrium.

The second law traces its origin to French physicist Sadi Carnot's 1824 paper Reflections on the Motive Power of Fire, which presented the view that motive power (work) is due to the fall of caloric (heat) from a hot to cold body (working substance). In simple terms, the second law is an expression of the fact that over time, differences in temperature, pressure, and density tend to even out in a physical system that is isolated from the outside world.

Note

- second law of thermodynamics : 열역학 제2법칙
- universal law : 보편적인 법칙
- tend to : ~ 하고자 하다

Rankine cycle

The rankine cycle is the basic vapor heat-engine cycle. The apparatus required consist of boiler, an engine or turbine, a condenser and a boiler feed pump. Operation is as follows;

Vapor at high pressure flows into the turbine or engine where it expands with the production of work, the expanded vapor enters the condenser where heat is removed condensing it to a liquid, the liquid is pumped into the boiler, and is there evaporated and returned to the turbine.

Note

- heat engine : 열기관
- reciprocating internal combustion engine : 왕복식 내연기관
- condenser : 응축기

05 Dynamics

Dimension

Physical quantity require quantitative descriptions when solving an engineering problem. Density is one such physical quantity. It is measure of the mass contained in a unit volume. There are nine quantities that are considered to be fundamental dimensions: length, mass, time, temperature, amount of a substance, electric current, luminous intensity, plane angle and solid angle. The dimensions of all other quantities can be expressed in terms of the fundamental dimension

Note

- physical quantity : 물리량
- density : 밀도
- measure : 측정량
- electric current : 전류
- amount of substance = number of mole : 몰 수
- electric current : 전류
- luminous intensity : 광도
- plane angle : 평면각
- solid angle : 입체각
- in terms of ~ : ~에 의하여
- dimension : 차원

The international system of units(SI)

In October 1960, the eleventh General (International) Conference on Weights and Measures redefined some of the original metric units and expanded the system to include other physical and engineering units. This expanded system is called **The international system of units.**

SI consists of seven **base units**, two **supplementary units** and derived units.

Base and supplementary units are defined as:

Metre(m) : The Metre is the length equal to 1,650,763.73 wave lengths in vacuum of the radiation corresponding to the transition between the levels $2p_{10}$ and $5d_5$ of the krypton 86 atom.

Kilogram(kg) : The Kilogram is the unit of mass; it is equal to the mass of the international prototype of the kilogram.

Second(s) : The Second is the duration of 9,192,631,770 periods of the radiation corresponding to the transition between the two hyperfine levels of the ground state of the cesium 133 atom.

Ampere(A) : The Ampere is that constant current which, if maintained in two straight parallel conductors of infinite length, of negligible cross section, and placed 1 meter apart in vacuum, would produce between these conductors a force equal to 2×10^{-7} newton per metre of length.

Kelvin(K) : The Kelvin, units of thermodynamic temperature, is the fraction 1/273.16 of the thermodynamic temperature of the triple point of water.

Mole(mol) : The Mole is the amount of substance of a system which contains as many elementary entities as there are atoms in 0.012 kilogram of carbon 12. (When the mole is used, the elementary entities must be specified and may be atoms, molecules, ions, electrons, other particles, or specified groups of such particles.)

Candela(cd) : The Candela is the luminous intensity, in the perpendicular direction, of a surface of 1/600,000 square metre of a blackbody at the temperature of freezing platinum under a pressure of 101,325 newtons per square metre.

Radian(rad) : The unit of measure of a plane angle with its vertex at the center of a circle and subtended by an equal in length to the radius.

Steradian(sr) : The unit of measure of a solid angle with its vertex at the center of a sphere and enclosing an area of the spherical surface equal to that of a square with sides equal in length to the radius.

[Table 5-1] SI prefixes

multiplication factor	prefix	symbol
10^{12}	tera	T
10^{9}	giga	G
10^{6}	mega	M
10^{3}	kilo	k
10^{-2}	centi	c
10^{-3}	milli	m
10^{-6}	micro	μ
10^{-9}	nano	n
10^{-12}	pico	p

Note

- base unit : 기본단위
- supplementary unis : 보조단위
- vacuum : 진공
- radiation : (빛, 열 등의) 방사 복사, 발열
- mass : 질량
- international prototype : 국제원기
- hyperfine : 초미세의
- cesium : 세슘(원소기호 Cs)
- thermodynamic : 열역학
- luminous intensity : 광도
- solid angle : 입체각
- vertex : 정점
- sphere : 구, 구형
- radius : 반지름, radii(복수)

Vector and scalar

In science, mathematics and engineering, we distinguish two important quantity: scalars and vectors. A scalar is simply a real number or quantity that has magnitude. For example, length, temperature, and blood pressure are represented by number such as 80 m, 20℃, and the systolic-diastolic ratio 120/80.

A vector, on the other hand, is usually described as a quantity that has both magnitude and direction. Geometrically, a vector is represented by a directed line segment- that is by an arrow-and is written either as a boldface symbol **v** or as $\vec{v}$ or . $\overrightarrow{AB}$

Note

- vector : 벡터
- scalar : 스칼라
- blood pressure : 혈압
- systolic-diastolic ratio : 최대-최소혈압
- friction : 마찰
- boldface : 볼드체

Newton's law

Law Ⅰ. A particle remains at rest or continues to move in a straight line with a constant velocity if there is no unbalanced force acting on it

Law Ⅱ. The acceleration of a particle is proportional to the resultant force acting on it and is in the direction of this force

Law Ⅲ. The forces of action and reaction between interacting bodies are equal in magnitude, opposite in direction, and collinear.

Note

- Newton's law : 뉴턴의 법칙
- particle : 입자
- velocity : 속도
- acceleration : 가속도
- action : 작용
- reaction : 반작용
- interacting : 상호작용하는

Newton's second law forms the basis for most of the analysis in dynamics. For a particle of mass m subject to a resultant force F, the law may be stated as

$$\vec{F} = m\vec{a}$$

where $\vec{a}$ is the resulting acceleration measured in a non-accelerating frame of reference. Newton's first law is a consequence of the second law since there is no acceleration when the force is zero, and the particle is either at rest or is moving with constant velocity.

The third law constitutes the principle of action and reaction with which we should be thoroughly familiar from our work in statics

Note

- Newton's second law : 뉴턴의 제 2법칙
- analysis : 해석, 분석
- dynamics : 동역학
- subject to ~ : ~를 받는, ~를 당하는
- consequence : 결과
- constitute : 구성하다
- action and reaction : 작용과 반작용
- statics : 정역학

cf. 형용사의 뒤에 s가 붙으면 그 형용사의 의미에 대한 과목의 이름이 된다.

ex. dynamic(역동적인) ↔ dynamics(동역학)
static(정적인) ↔ statics(정역학)
Economic(경제적인) ↔ economics (경제학)
statistic(통계의) ↔ statistics(통계학)
thermodynamic(열역학적인) ↔ thermonamics(열역학)

\ Relative motion

It is not always possible or convenient to use a fixed set of axes for the observation of motion and there are many engineering problems for which the analysis of motion is simplified by using measurements made with respect to a moving reference system. These measurements, when combined with the absolute motion of the moving coordinate system, permit us to determine the absolute motion in question. This approach is known as a relative-motion analysis.

Note

- relative motion : 상대운동
- observation : 관찰
- with respect to ~ : ~에 관하여
- determine : 결정하다

Kinematics

Kinematics is that branch of dynamics which describes the motion of bodies without reference to the force which either cause the motion or are generated as a result of the motion. Kinematics is often referred to as the "geometry of motion". The design of cams, gears, linkage, and other machine elements to control of produces certain desired motions and the calculation of flight trajectories for aircraft, rockets, and spacecraft are a few examples for kinematics problems which engage the attention of engineer

Note

- kinematics : 운동학
- geometry : 기하학
- trajectory : 궤적
- spacecraft : 우주선

Kinetics

Kinetics is the study of the relations between unbalanced forces and the changes in motion that they produce. There are three general approaches to the solution of kinetics problems

(a) direct application of newton's second law(called the force-mass-acceleration method)
(b) use of work and energy principles
(c) solution by impulse and momentum method

Note

• kinetics : 동역학	• study : 연구
• impulse : 충격량	• momentum : 운동량

Work and Kinetic energy

The work is the product of the magnitude of the force acting in the direction of motion and the displacement. Therefore if no motion takes place in the direction of the force, then mechanically no works has been made

The k inetic energy T of the particle is defined as ;

$$T = \frac{1}{2}mv^2$$

and is the total work which must be done on the particle to bring it from a state of rest to a velocity v. Kinetic energy T is a scalar quantity with the unit of N · m or joule(J). Kinetic energy is always positive, regardless of the direction of the velocity

Note

• work : 일	• kinetic energy : 운동에너지
• take place : 발생하다	• particle : 입자

Potential energy

The gravitational potential energy V_g of the particle is defined as the work mgh done against the gravitational field to elevate the particle a distance h above some arbitrary reference plane where V_g is zero. Thus we write the potential energy as ;

$$V_g = mgh$$

The elastic potential energy is found in the deformation of an elastic body, such as a spring. The work which is done on the spring to deform it is stored in the spring and is called the elastic potential energy V_e. This energy is recoverable in the form of work done by the spring on the body attached to its movable end during the release of the spring's deformation. We define the elastic potential energy of the spring as the work done on it to deform it an amount x and we have ;

$$V_e = \frac{1}{2}kx^2$$

Where, k is spring constant.

Note

- gravitational potential energy : 중력위치에너지
- gravitational field : 중력장
- elastic body : 탄성체
- elastic potential energy : 탄성위치에너지
- deform : ~를 변형시키다
- be stored : 저장되다
- deformation : 변형
- spring constant : 스프링 상수

Energy conservation

Total mechanical energy of the particle and its attached linear spring is ;

$$E = T + V_g + V_e$$

When E is constant, we see that transfers of energy between kinetic and potential may take place as long as the total mechanical energy does not change.

Note

- energy conservation : 에너지 보존
- mechanical energy : 역학적인

Translation

Translation is defined as any motion in which every line in the body remains parallel to its original position at all time. In translation there is no rotation of any line in the body. In rectilinear translation, part (a) of figure. 5-1, all point in the body move in parallel straight lines. In curvilinear translation, all points move on congruent curves. We note that in each of the two cases of translation, the motion of body is completely specified by the motion of any point in the body, since all points have the same motion.

Note

- translation : 병진운동
- be defined as ~ : ~로써 정의되다
- congruent curve : 합동(일치)인 곡선
- parallel : 평행의

Rotation

Rotation about a fixed axis, part (c) of figure. 5-1 is the angular motion about the axis. It follows that all particles move in circular paths about the axis of rotation, and all lines in the body which are perpendicular to axis of rotation(including those that do not pass through the axis) rotate through the same angle in the same time.

Note

- angular motion : 각운동
- axis of rotation : 회전 축
- perpendicular to ∼ : ∼에 수직인

(a) Rectilinear	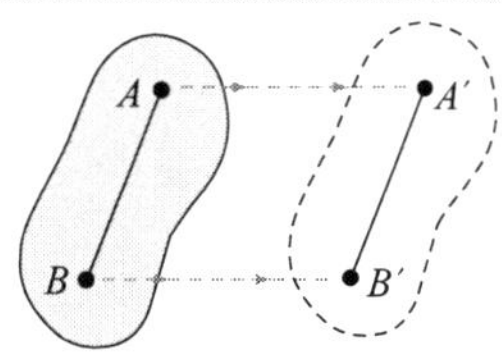	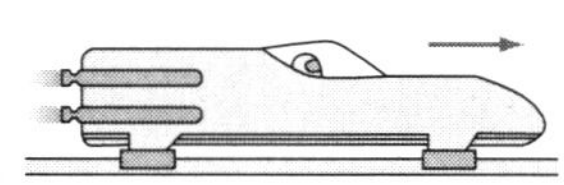 Rocket test sled
(b) Curvilinear translation	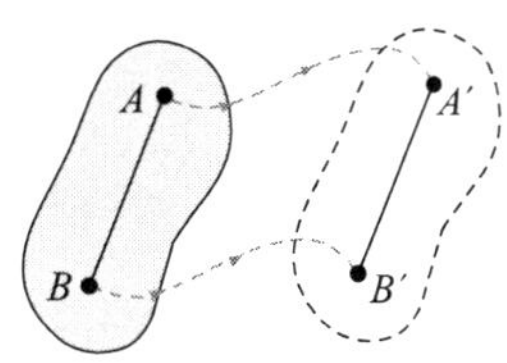	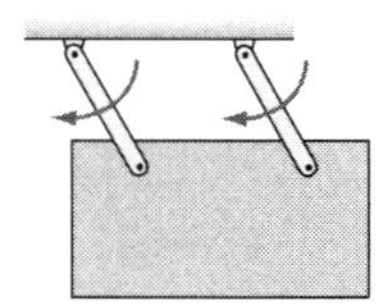 Parallel−link swinging
(c) Fixed-axis rotation	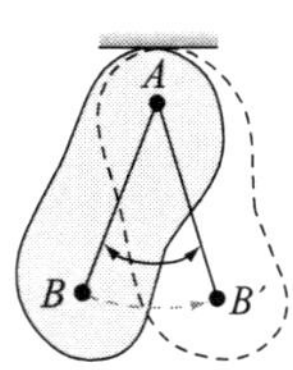	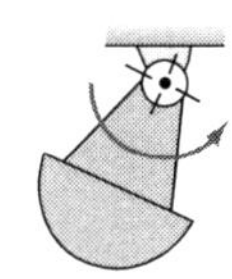 Compound pendulum
(d) General plane motion	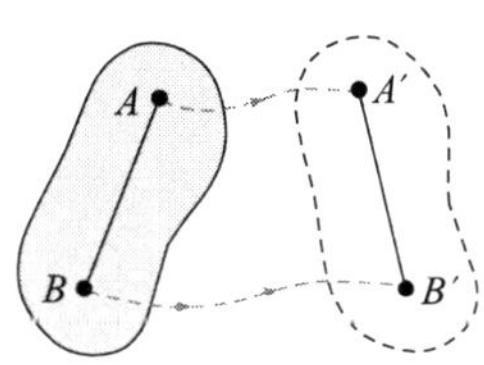	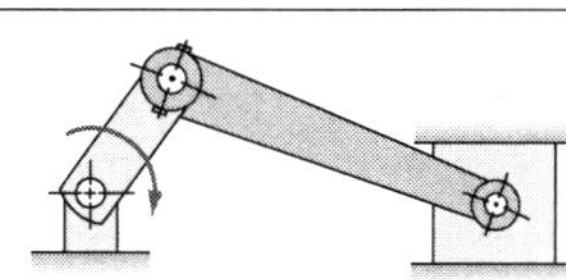 Connecting rod in a reciprocating engine

[Figure 5−1] Type of rigid−body plane motion and its example

06 Mechanical Drawing

Technical graphics

Technical graphics is a real and complete language used in the design process for ;

1. Visualization
2. Communication
3. Documentation

A drawing is a graphical representation of objects and structures and is done using freehand, mechanical, or computer methods. A drawing serves as a graphic model or representation of a real object or idea.

Note

- mechanical drawing : 기계제도
- visualization : 도면화, 시각화, 구상화
- communication : 정보전달
- documentation : 문서화

Visualization

A powerful tool for design engineers is the ability to see in their minds the solution to problems. Visualization is the ability to mentally picture things that do not exist. Design engineers with good visualization ability not only are able to picture things in their minds, but are able to control that mental image allowing them to move around the image, change the form, look inside, and other movement as if they were holding the object in their hands. Some of the greatest engineers and scientists throughout history have had powerful visualization ability, such as Albert Einstein, Leonardo Da Vinci, and Thomas Edison.

Note

- Albert Einstein : 1879~1955. 이론물리학자.
  1905년 특수상대성이론을 발표하여 뉴턴 역학을 뒤흔들었으며, 1914~1916년 일반상대성 이론을 완성. 미국 망명 후 원폭제조에 관여, 전후에는 핵무기 금지, 평화운동에 헌신
- Thomas Edison : 1847~1931 미국발명가
  1876년 축음기, 1878 백열전구, 1893, 영화발명
  전생애 동안 특허 1300여개로 발명왕으로 불리운다
- Leonardo Da Vinci : 1452~1519
  르네상스 전성기의 예술가, 과학자, 해부생리학, 조류의 비행원리 연구

The history of graphics communications

Drawing is so old that its history is virtually that of humanity, and it closely parallels human technological progress. Drawing may be called a "universal language". It is the natural method for humans to communicate visual images of the mind. As far back as 12,000 B.C., cave drawings recorded and depicted aspects of the prehistoric human experience. Drawings to communicate technical ideas may even predate the advent of written language.

Note

- universal language : 만국언어, 세계공통언어

Technical sketching

There are three methods of creating technical drawings: freehand, mechanical, and digital. Technical sketching is the process of producing a rough, preliminary drawing representing the main features of a product or structure. Such sketches have traditionally been done freehand: today, CAD systems can also be used. A technical sketch is generally less finished, less structured or restricted, and it takes less time than other types of freehand illustrations.

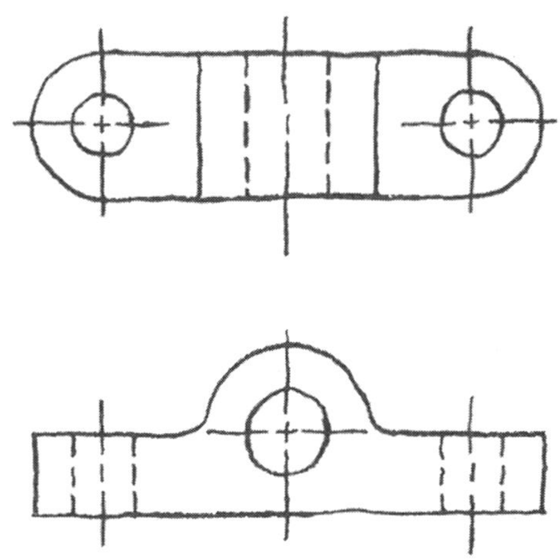

[Figure 6-1] A rough technical sketch can be made to capture a design idea quickly.

Freehand sketching tools

Normally, tools used for sketching should be readily available and usable anywhere: pencil, and eraser. Although variations on these tools are numerous and sophisticated, the goal of technical sketching is simplification. Just a few pencils, an eraser, and sheets of paper should be all that is needed. Many a great design idea was born on the back of a napkin with a wooden pencil! Grid papers use in certain types of pictorial sketch.

Note

- Grid paper : 방안지

How to sketching long lines

Very long lines can sometimes be more difficult to draw. One techniques is to use the edge of the paper as a guide for your hand. Another technique is to mark equal distances from the edge of the paper using a marked scrap of paper as a guide. The marks are then used as a guide to produce the line

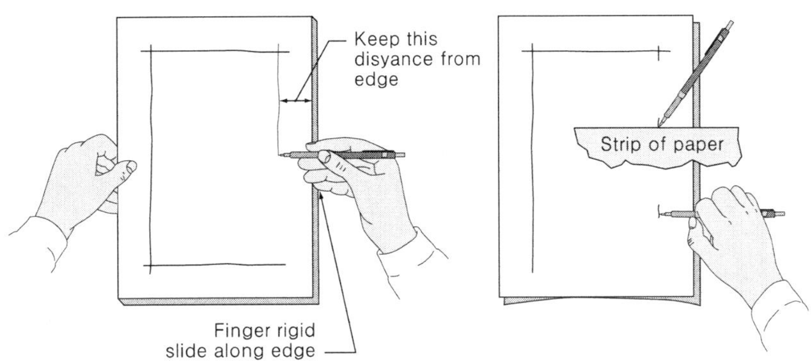

[Figure 6-2] Sketching long line

Six principal views

A multiview drawing of an object will produce six views, called regular or principal views. The object is viewed from six mutually perpendicular viewpoints. The six views are called front, top, bottom, rear, right side, and left side.

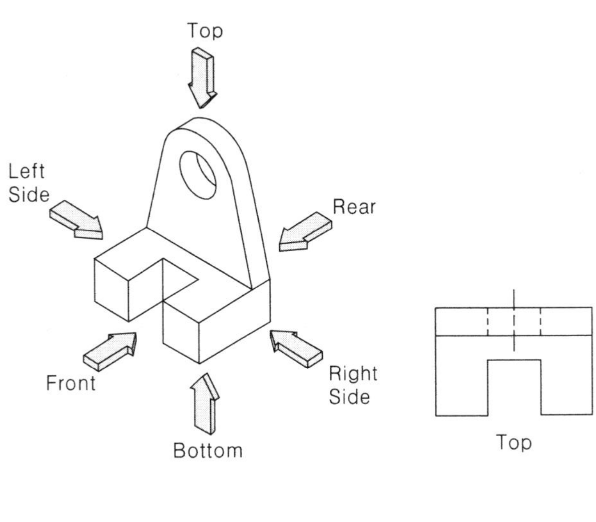

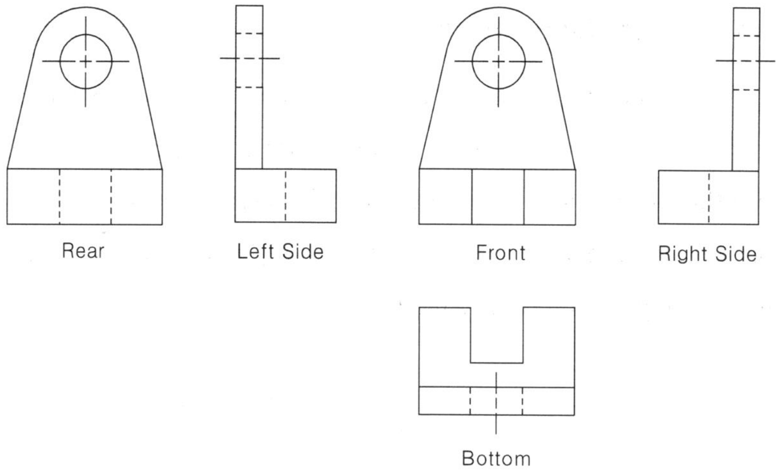

[Figure 6-3] Six principal views

Note

- front View : F.V, 정면도
- top View : T.V, 평면도
- right side view : R.S.V 우측면도
- left side view : L.S.V 좌측면도
- rear view : 배면도
- bottom view : 저면도

Circle definitions

A circle is a single-curved plane with all points at an equal distance from a point called the center. The important features of a circle are shown (Figure 6-4).

- **Center** : The midpoint of the circle.
- **Circumference** : The distance all the way around the circle.
- **Radius** : A line joining the center to any point on the circumference.
- **Chord** : A strait line joining any two points on the circumference.
- **Diameter** : A chord that passes through the center. The diameter is equal to twice the radius.

- **Secant** : A straight line that goes through a circle but not through the center.
- **Arc** : A continuous segment of the circle.
- **Semicircle** : An arc measuring one-half the circumference of the circle.
- **Minor arc** : An arc that is less than a semicircle.
- **Major arc** : An arc that is greater than a semicircle.
- **Central angle** : An angle formed by two radii.
- **Sector** : An area bounded by two radii and an arc, usually a minor arc.
- **Quadrant** : A sector equal to one-fourth the area of the circle. The radii bounding a quadrant are at a right angle to each other.
- **Segment** : An area bounded by a chord and a minor arc. The chord does not go through the center.
- **Tangent** : A line that touches the circle at one and only one point.
- **Concentric circles** : Circles of unequal radii that have the same center point.
- **Eccentric circles** : Circles of unequal radii that have different centers, and one circle is inside the other.
- **Circumscribed circle** : A circle drawn outside of a polygon such that each vertex of the polygon is a point on the circle.
- **Inscribed circle** : A circle drawn inside a polygon such that all sides of the polygon are tangent to the circle.

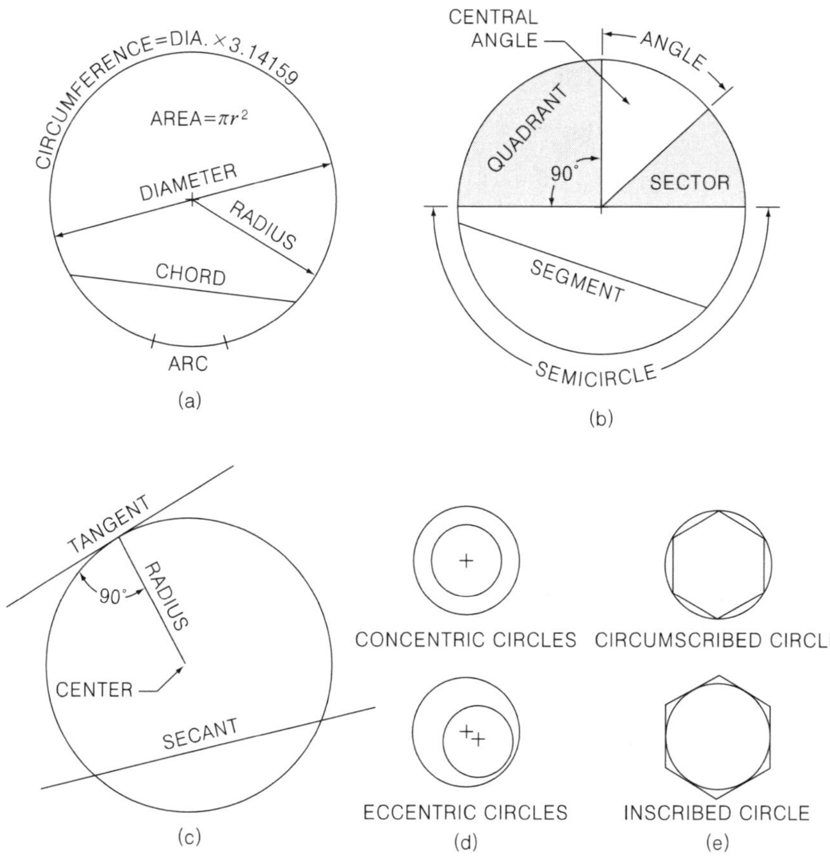

(a) (b) (c) (d) (e)

[Figure 6-4] Circle definitions

Regular polygons

A polygon is a multisided plane of any number of sides. If the sides of the polygon are equal in length, the polygon is called a regular polygon. Regular polygons can be inscribed in circles. To determine the angle inside a polygon, use the equation $S = (n-2) \times 180$ degrees. Where n equals the number of sides. Regular polygons are grouped *by the number of sides* (Figure 6-5)

- **Triangle(equilateral)** : Three equal sides and angles.
- **Square** : Four equal sides and angles.
- **Pentagon** : Five equal sides and angles.

- **Hexagon** : Six equal sides and angles.
- **Heptagon** : Seven equal sides and angles.
- **Octagon** : Eight equal sides and angles.
- **Nonagon** : Nine equal sides and angles.
- **Decagon** : Ten equal sides and angles.
- **Dodecagon** : Twelve equal sides and angles.
- **Icosagon** : Twenty equal sides and angles.

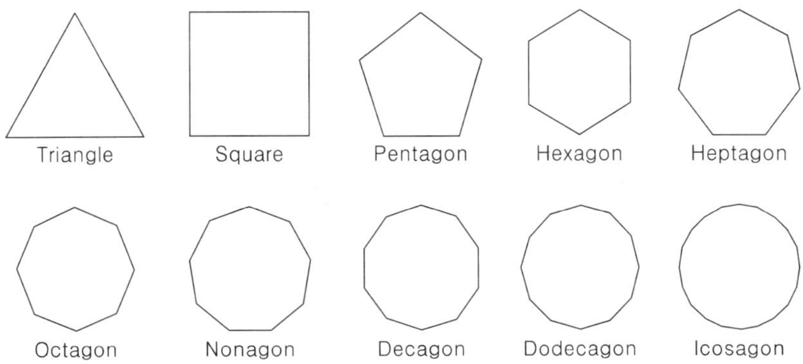

[Figure 6-5] classification of polygons

Triangles

A triangle is a polygon with three sides. The sum of the interior angles equals 180 degrees. The vertex is the point at which two of the sides meet. Triangles are named according to their angles (right. obtuse. acute) or the number of equal sides(Figure 6-6).

- **Equilateral triangle** : Three equal sides and three equal interior angles of 60 degrees.
- **Isosceles triangle** : At least two equal sides.
- **Scalene triangle** : No equal sides or angles.
- **Right triangle** : Two sides that from a right angle (90 degrees), and the square of the hypotenuse is equal to the sum of the squares of the two sides (Pythagoras Theorem).

• **Acute triangle** : No angle larger than 90 degrees.

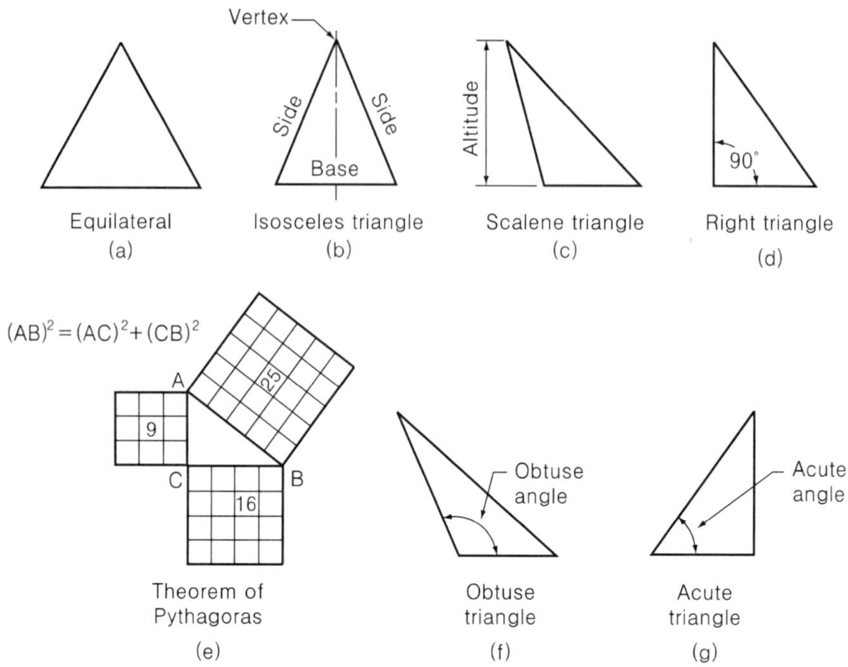

[Figure 6-6] Classification of triangle

Constructing a Triangles

Step 1. If you are given the length of the sides, A B, and C, but not the internal angles, draw one of the sides, such as A.

Step 2. Draw an arc with a radius equal to the length of B, using one end of line A as the center.

Step 3. Draw sides B and C from the endpoints of line A to the point of intersection of the two arcs (Figure 6-7).

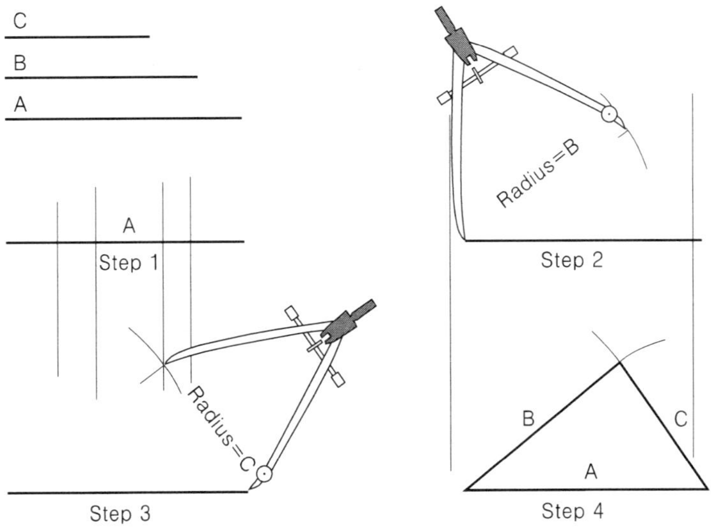

[Figure 6-7] Constructing a triangles, given the length of its sides

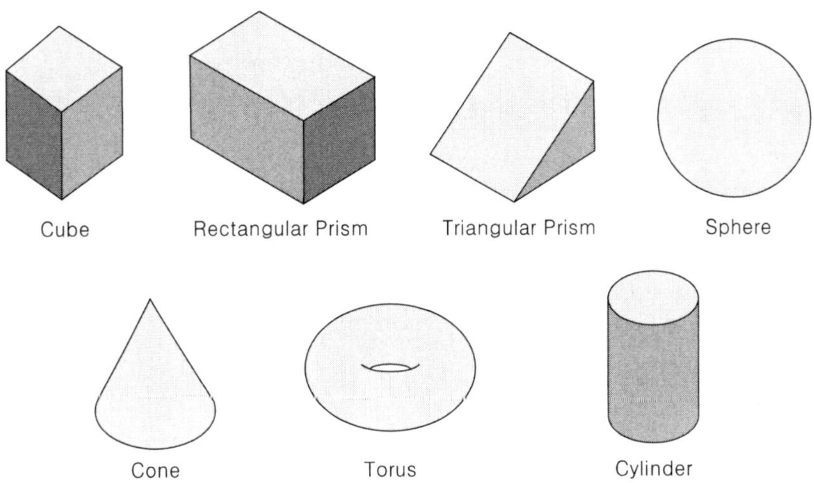

[Figure 6-8] Common geometric primitives

Tolerancing

Tolerances are used to control the variation that exists on all manufactured parts. Toleranced dimensions control the amount of variation on each part of an assembly. The amount each part is allowed to vary depends on the function of the part and of the assembly. The more accuracy needed in the machined part, the higher the manufacturing cost. Therefore, tolerances must be specified in such a way that a product functions as it should at a cost that is reasonable.

The tolerance of 4.650 ± 0.003 means that the final measurement of the machined part can be anywhere from 4.653 to 4.647 and the part would still be acceptable. The **upper limit** (largest value) for the part is 4.653, the **lower limit**(smallest value) is 4.647, and the tolerance is 0.006.

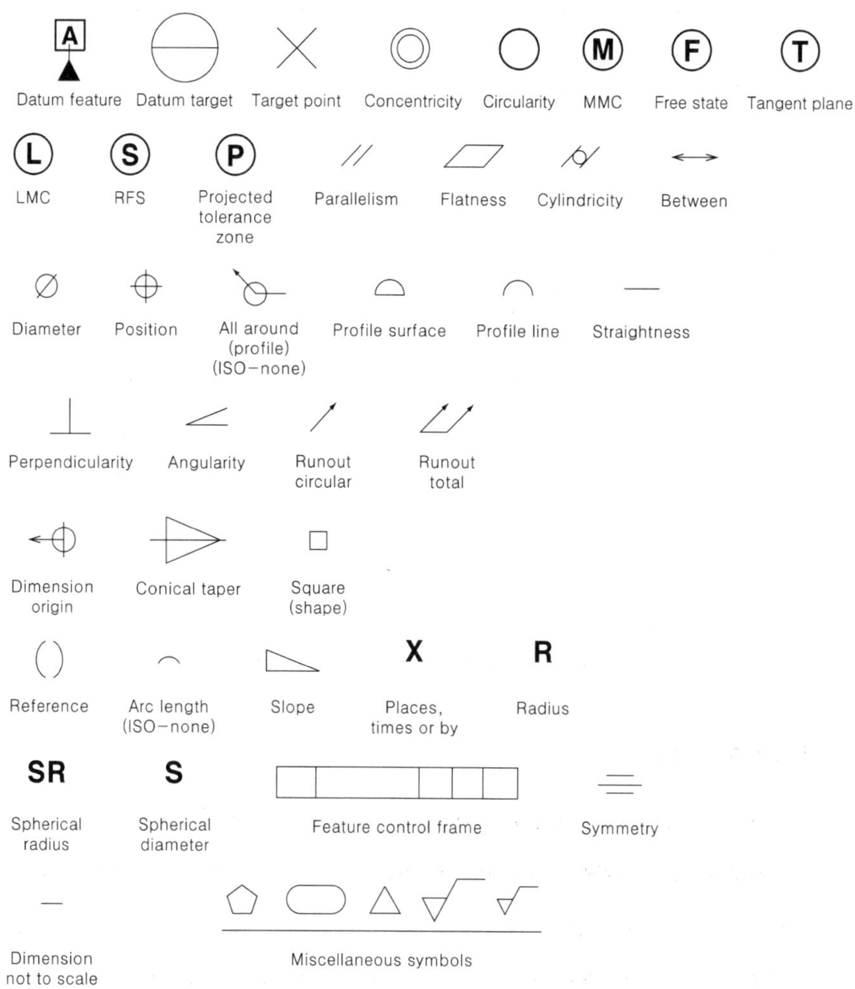

[Figure 6–9] Dimensioning and tolerancing symbols

Note

- tolerance : 공차
- lower limit : 최소허용치수
- datum target : 데이텀 타켓 기입틀
- circularity : 진원도
- flatness : 평면도
- profile surface : 면의 윤곽도
- perpendicularity : 직각도
- runout circular : 원주 흔들림
- symmertry : 대칭도
- miscellsneous symbols : 여러 가지 기호들
- upper limit : 최대허용차수
- datum feature : 데이텀 형체
- concentricity : 동축도, 동심도
- parallelism : 평행도
- cylindricity : 원통도
- straightness : 직진도
- angularity : 경사도
- roundout total : 온 흔들림

Thread Definitions

- **Axis** : The longitudinal center line that passes through the screw thread cylinder.
- **Chamfer** : The angular relief at the last thread; allows easier engagement with the mating part.
- **Crest** : The peak or top surface of a screw thread.
- **Depth** : The distance between the crest and the root of a thread, measured normal to the axis.
- **Die** : A tool used to form external threads.
- **External thread** : The screw thread on the outside of a cylindrical or conical surface.
- **Internal thread** : The screw thread on the inside of a cylindrical or conical surface.
- **Lead** : The distance a screw will travel when turned 360 degree, or one revolution.
- **Major diameter** : The largest diameter on an internal thread.
- **Minor diameter** : The smallest diameter on an internal thread.
- **Pitch** : The distance between corresponding points on adjacent thread forms, measured parallel to the axis. The pitch is equal to the number 1 divided by the number of threads per inch.

For example, a screw threads with 12 threads per inch would have a pitch 1/12.

- **Pitch diameter** : The diameter of an imaginary cylinder that is located equidistant between the major and minor diameters.
- **Root** : The bottom of a screw thread cut into a cylinder.
- **Screw thread** : A continuous and projecting helical ridge on a cylindrical or conical surface.
- **Side** : The screw thread surface that connects the crest and root.
- **Tap** : A tool used to make threads in holes.
- **Tap drill** : A drill bit used to make a hole in metal before tapping an internal thread.
- **Thread angle** : The angle between the surface of two adjacent threads.
- **Thread form** : The profile or shape of a thread cut into a cylinder.
- **Thread series** : The number of threads per inch for a given diameter.
- **Thread per inch** : The number of threads in one inch, measured axially(parallel to the axis); the reciprocal of the pitch.

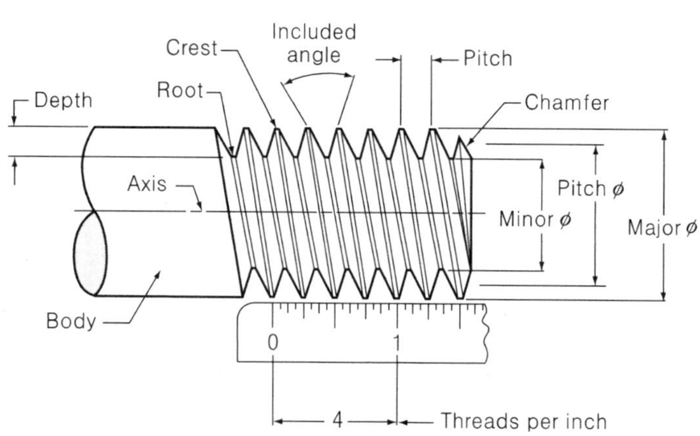

External threads

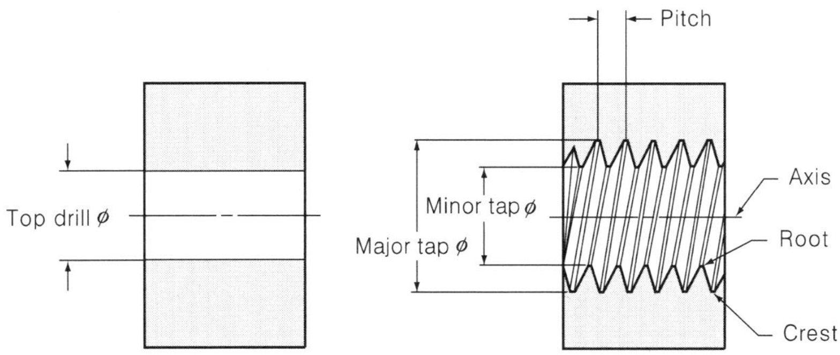

Internal threads

[Figures 6-10] Thread Definitions

Note

- chamfer : 모서리를 떼낸 사면(chamfering, 모떼기)
- crest : 나사산봉우리
- crest : 나사산 봉우리
- depth : 나사깊이
- external thread : 수나사
- internal thread : 암나사
- lead : 리드
- pitch : 피치
- root : 골밑
- screw thread : 나사산
- thread angle : 나사산의 각도

Bolts, Studs, and Screws

The five general types of standard threaded fasteners are as follow.

- **Bolt** : A mechanical threaded device with a head on one end and threads on the other end. Bolts are paired with nuts. A nut is a mechanical threaded device used on the end of a bolt, stud, or machine screw.
- **Stud** : A rod that is threaded both ends and joins two mating parts. A nut may be used on one end.
- **Cap screw** : A mechanical threaded device with a head on one end and threads on the other end. Cap screws join two mating parts and have longer threaded than bolts. Also, cap screw can be made with slotted heads.

- **Machine screw** : A mechanical threaded device with a head on one end and threads on the other end. The threaded end may screw into a mating part, or may be used with a nut. A machine screw is similar to a cap screw, but it is normally smaller.
- **Set screw** : A mechanical threaded device with or without a head, used to prevent rotation or movement between parts, such as a shaft and a collar. Set screws have different types of points and heads for different applications.

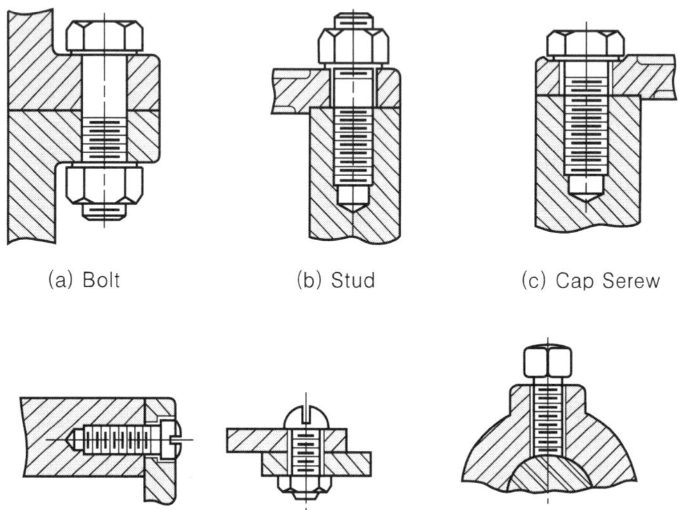

(a) Bolt (b) Stud (c) Cap Serew

(d) Machine (e) Set Screw

[Figure 6-11] General types of fasteners

The five general types of fasteners are bolts, studs, cap screws, machine screws and set screws.

Note

- fasten : 체결하다
  fastner : 볼트 등 체결구
- stud : 스텃볼트
- set screw : 고정나사, 세트나사

Springs

Springs are mechanical device that expand or contract due to pressure or force. Many types of springs have a helix form. Spring classifications are compression, tension, extension, flat, and constant force. Springs are further classified by the material, gage, spring index, helix direction, and type of end.

Springs can be represented on technical drawings either in detail or in schematic form (Figure 6-12). The schematic representation is drawn by laying out the dimeter, the free length, and then the diagonals to represent the number of coils. With detailed drawings for springs, phantom lines can be used to save time. Straight lines are used instead of helical curves.

Springs are specified by giving the inside and outside diameters, wire gage, type of ends, material, finish, number of coils, free and compressed lengths.

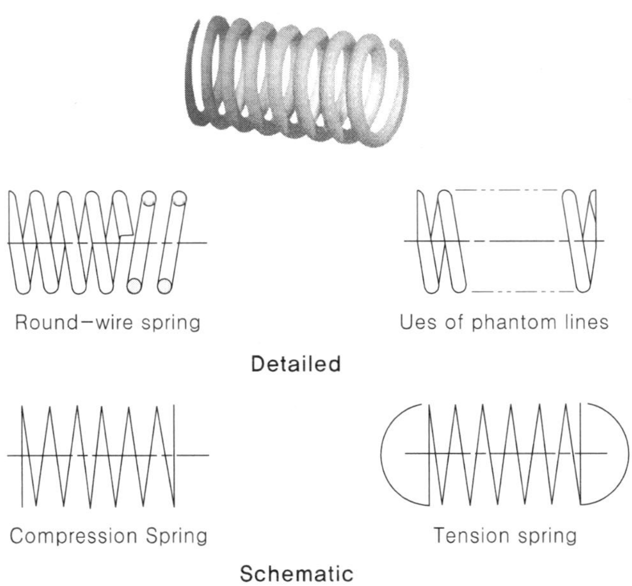

[Figure 6-12] Springs are represented either in detail or in schematic form

Note

- helix form : 나선형
- diagonal : 대각선
- schematic : 개략도, 약도
- detailed drawing : 상세도

Cams

A cam is a mechanical device with a surface or groove that controls the motion of a second part, called a follower (Figure 6-13). Cams convert rotary motion to linear motion. The timing and manner of the movement to be created by the cam are the main elements in designing cams. For example, a cam might be used to raise a follower rapidly, keep the follower stationary, then allow the follower to fall slowly to its beginning position. This is all done during the 360-degree revolution of the cam and is called the displacement. Cams on a cam shaft are used in automobile engines to control the opening and closing of the valves.

Note

- follower : 캠과 같이 돌아가는 종동자
- rotary motion : 회전운동
- linear motion : 직선운동
- revolution : 회전(운동)

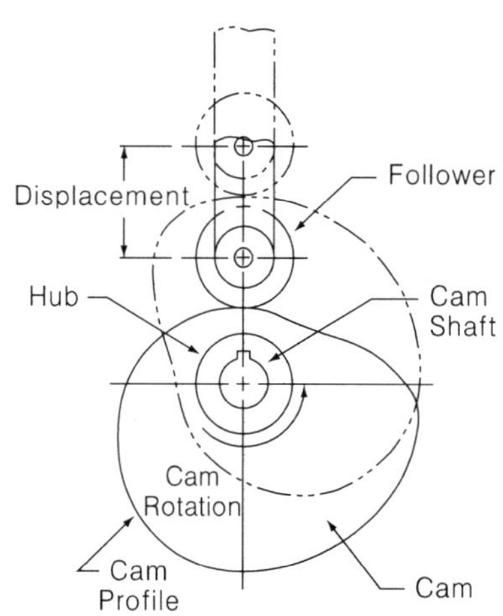

[Figure 6-13] Cam and follower

Types of cams

Face cams are flat disks, and the shaft controls the movement of the follower. Groove cams are flat disks with grooves cut in them, and the grooves control the movement of the follower. Cylindrical cams are cylinders with grooves cut in them, and the grooves control the movement of the follower.

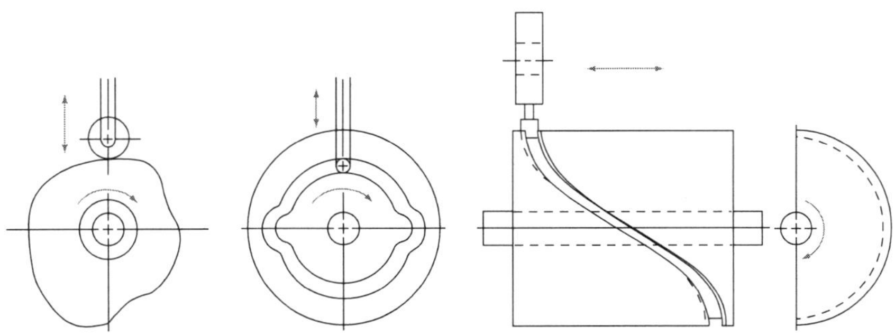

[Figure 6-14] types of cams

Note

- face cam : 정면캠
- groove cam : 홈캠
- cylindrical cam : 원통캠

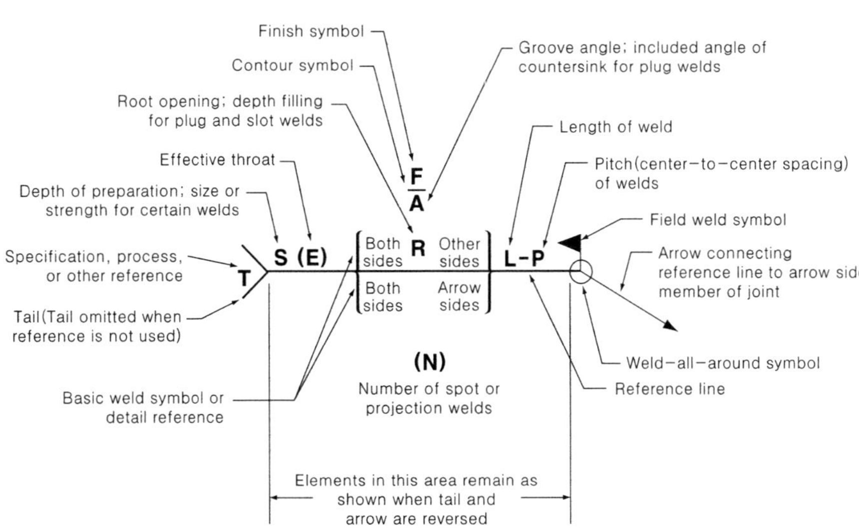

[Figure 6-15] Weld symbols

07 Cad/Cam

Computer Aided Design(CAD)

CAD originally meant Computer-Aided Drafting or Designing because of its original use as a replacement for traditional drafting. Now, CAD usually means Computer Aided Design to reflect the fact that modern CAD tools do more than just drafting.

CAD is used to design, develop and optimize products, which can be goods used by end consumers or intermediate goods used in other products. CAD is also extensively used in the design of tools and machinery used in the manufacture of components, and in the drafting and design of all types of buildings, from small residential types (houses) to the largest commercial and industrial structures (hospitals and factories).

CAD is mainly used for detailed engineering of 3D models and/or 2D drawings of physical components, but it is also used throughout the engineering process from conceptual design and layout of products, through strength and dynamic analysis of assemblies to definition of manufacturing methods of components.

Note

- drafting : 제도
- reflect : 반영하다
- optimize : 최적화하다
- extensively : 널리
- machinery : 기계
- residential : 주거의
- traditional : 전통적인
- end consumers : 최종소비자
- intermediate goods : 중간재
- tool : 공구
- manufacture : 제조
- commercial : 상업용의

- industrial : 산업용의
- 3D : 3 dimensional, 3차원의
- layout : 배치, 설계

Computer Aided Manufacturing(CAM)

Computer-aided manufacturing (CAM) is the use of computer-based software tools that assist engineers and machinists in manufacturing or prototyping product components. CAM is a programming tool that allows you to make 3D models using computer-aided design (CAD). CAM was first used in 1971 for car body design and tooling.

Traditionally, CAM has been considered as an NC programming tool wherein 3D models of components generated in CAD software are used to generate CNC code to drive numerical controlled machine tools.

Although this remains the most common CAM function, CAM functions have expanded to integrate CAM more fully with CAD/CAM/CAE solutions.

Note

- manufacturing : 생산
- prototype : 시작품
- component : 구성요소, 부품
- 3D : 3 dimensional, 3차원의
- traditionally : 전통적으로
- NC : (=numerical control)수치제어
- where in : 그 중에
- CNC : Computer Numeried Control : 컴퓨터 수치제어
- common : 일반적인
- expanded : 확장하다
- integrate : 통합하다
- CAE : Computer-Aided Engineering : 컴퓨터응용공학

Numerical Control(NC)

Numerical control or numerically controlled (NC) machine tools are machines that are automatically operated by commands that are received by their processing units. NC machines were first developed soon after World War II and made it possible for large quantities of the desired components to be very precisely and efficiently produced (machined) in a reliable repetitive manner. These early machines were often fed instructions which were punched onto paper tape or punch cards. In the 1960s, NC machines largely gave way to CNC, or computer numerical control, machines. (GE had its NC 550 workhorse for many years until they came out with their first CNC (model 1050) in August 1974.)

Note

- numerical control : 수치제어
- command : 명령어
- processing unit : 처리장치
- precisely : 정밀하게
- efficiently : 효율적으로
- reliable : 신뢰할 수 있는
- repetitive : 되풀이하는, 반복성의
- GE : General Electric, 미국의 종합전기 메이커

CNC machine

In a production environment, a series of CNC machines may be combined into one station, commonly called a "cell", to progressively machine a part requiring several operations. CNC machines today are controlled directly from files created by CAM software packages, so that a part or assembly can go directly from design to manufacturing without the need of producing a drafted paper drawing of the manufactured component. In a sense, the CNC machines represent a special segment of industrial robot systems, as they are programmable to perform many kinds of machining operations (within their designed physical limits, like other robotic systems). CNC machines can run over night and over weekends without operator intervention.

Note

- production : 생산
- commonly : 일반적으로
- part : 부품
- assembly : 조립품
- a drafted paper drawing : 손으로 작성된 도면
- in a sense : 어떤 점에서, 어느 정도까지
- segment : 부분
- intervention : 간섭, 조정, 중재

[Figure 7-1] A CNC Turning Center

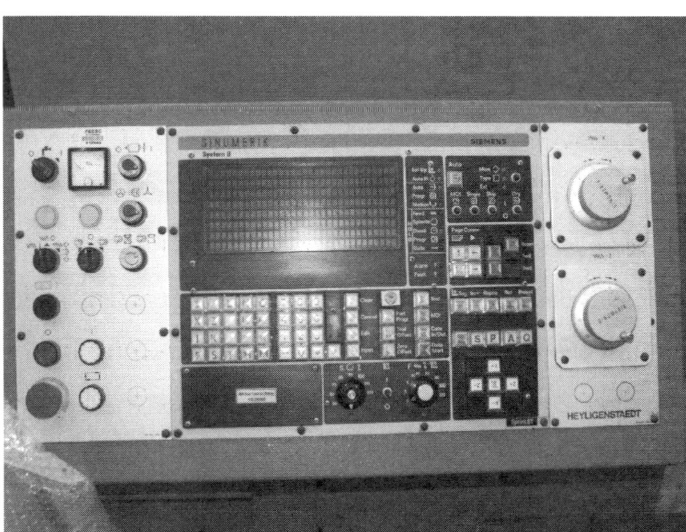

[Figure 7-2] CNC panel

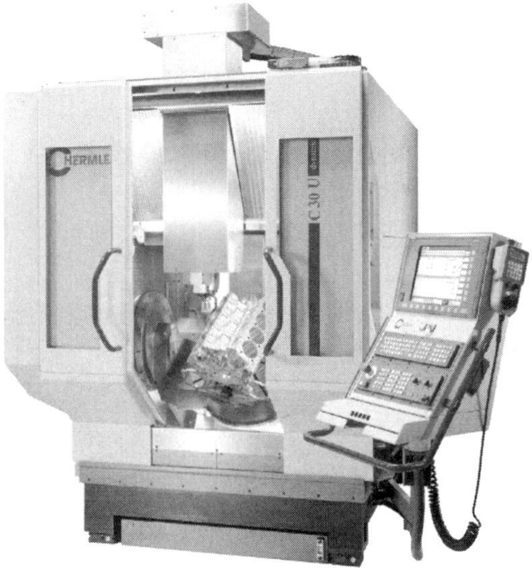

[Figure 7-3] A CNC Milling Machine

Reverse Engineering

Reverse engineering (RE) is the process of discovering the technological principles of a device, object or system through analysis of its structure, function and operation. It often involves taking something (e.g. a mechanical device, electronic component, or software program) apart and analyzing its workings in detail, usually to try to make a new device or program that does the same thing without copying anything from the original.

Reasons for reverse-engineering is as follows ;

Reverse engineering often is done because the documentation of a particular device has been lost (or was never written), and the person who built the thing is no longer available. Integrated circuits often seem to have been designed on obsolete, proprietary systems, which means that the only way to incorporate the functionality into new technology is to reverse-engineer the existing chip and then re-design it.

Note

- reverse engineering : 역공학
- discover : 발견하다
- device : 기구
- object : 사물
- apart : 분리하여, 분해하여
- in longer ~ : 더 이상 ~ 하지 않다.
- obsolete : 쓸모 없는, 뒤떨어진
- proprietary : 독점의

Pro/ENGINEER Wildfire 4.0

Pro/ENGINEER is an integrated 3D CAD/CAM/CAE solution for mechanical engineering and design which was created by Dr. Samuel P. Geisberg in the mid-1980's. It pioneered the CAD industry by introducing the concept of Parametric, Feature-based Solid Modeling. Models are driven parameters and intelligent features, rather than simple dimensional values.

Pro/ENGINEER outputs consist of solid model data for tooling and rapid prototyping, CNC manufacturing, and finite element analysis. A product and its entire Bill of Materials can be modeled accurately with fully associative engineering drawings, and revision control information. It is compatible with Windows and Unix-variants. All data is interchangeable between these platforms without conversion.

Note

- pioneer : 개척하다
- drive : 가동시키다
- consist of : ~으로 구성된
- prototype : 시작품
- finite element analysis : FEM, 유한요소해석법
- entire : 전체의
- associative : 조합하는
- revision : 개정
- Windows : Microsofr사의 마이크로 컴퓨터 운용시스템
- Unix : 미국 벨 연구소가 1973년 개발한 운영체제
- interchangeable : 교환할 수 있는
- compatible : 호환가능한
- without conversion : 변환 없이

CATIA V5

CATIA(Computer Aided Three Dimensional Interactive Application) is a multi-platform CAD/CAM/CAE commercial software suite developed by French company Dassault Systemes and marketed world-wide by IBM. The software was originally intended for the development of Dassault's Mirage fighter jet, but became a runaway success and was subsequently adopted by numerous well known companies world-wide, such as Boeing and IBM. The software was also famously used by architect Frank Gehry in his building of the Guggenheim Museum Bilbao. CATIA is written in the C++ programming language. CATIA is the corner stone of the Dassault Systems PLM software suite.

Note

- commercial : 상업적인
- intend : ~할 작정이다
- runaway : 결정적인, 일방적인
- subsequently : 계속해서
- be adopted by : ~에 의해 채용되다
- such as : ~과 같이
- architect : 건축가
- C++ : C 언어를 확장한 객체지향프로그래밍언어
- corner stone : 조석
- PLM : Product Lifecycle management, 제품수명주기관리

UG NX5

UGS NX is the commercial CAD/CAM/CAE PLM software suite developed by Siemens PLM Software. NX is widely used in the engineering industry, especially in the automotive and aerospace sectors. NX has some presence in the consumer goods design sector.

NX is a parametric solid / surface feature-based modeler. It uses the Parasolid geometric modeling kernel.

NX is commonly referred to as a 3D PLM software application. The product supports all stages of product development from conceptualization (CAID), to design (CAD), to analysis (CAE), to manufacturing (CAM). NX integrates these product lifecycle stages into an end-to-end process using concurrent engineering workflow, design-in-context and product data management tools that apply across all functional areas.

Note

- automotive : 자동차
- aerospace : 항공우주산업
- presence : 존재
- consumer goods : 소비재
- geometric : 기하학적
- kernel : 핵심
- conceptualization : 개념화
- CAID : Computer Aided Industrial Design
- end-to-end : 철저한, 샅샅이
- concurrent engineering : 동시공학
- context : 환경

Autocad

AutoCAD is a CAD software application for 2D and 3D design and drafting, developed and sold by Autodesk, Inc. Initially released in late 1982, AutoCAD was one of the first CAD programs to run on personal computers, and notably the IBM PC. Most CAD software at the time ran on graphics terminals connected to mainframe computers or mini-computers.

In earlier releases, AutoCAD used primitive entities-such as lines, polylines, circles, arcs, and text-as the foundation for more complex objects. Since the mid-1990s, AutoCAD has supported custom objects through its C++ API. Modern AutoCAD includes a full set of basic solid modeling and 3D tools, but lacks some of the more advanced capabilities of solid modeling applications.

Note

- design : 설계
- drafting : 제조
- Inc : Incorporated, 주식회사
- initially : 최초로
- release : 발표하다
- primitive : 원시의
- entity : 실체
- poly line : 여러점에 직선을 연결하는 것
- API : Application Programming Interface, 프로그램내에서 실행을 위해 특정 서브루틴에 연결을 제공하는 함수를 호출하는 것.

3D scanner

A 3D scanner is a device that analyzes a real-world object or environment to collect data on its shape and possibly its appearance (i.e. color). The collected data can then be used to construct digital, three dimensional models useful for a wide variety of applications. These devices are used extensively by the entertainment industry in the production of movies and video games. Other common applications of this technology include industrial design, reverse engineering and prototyping, computer vision and documentation of cultural artifacts. Many different technologies can be used to build these 3D scanning devices; each technology comes with its own limitations, advantages and costs. It should be remembered that many limitations in the kind of objects that can be digitized are still present: for example optical technologies encounter many difficulties with shiny, mirroring or transparent objects.

Note

- shape : 형상
- appearance : 외관
- entertainment industry : 오락산업
- reverse engineering : 역공학
- prototype : 시작품
- artifact : 인공물
- encounter : 부딪치다
- shiny : 빛이 나는
- transparent : 투명한

08 Concurrent Engineering

Concurrent Engineering(CE)

Concurrent engineering is a systematic approach to the integrated, concurrent design of products and their related process, including manufacture and support. This approach is intended to cause the developers, from the outset, to consider all elements of the product life cycle from concept through disposal, including quality, cost, schedule, and user requirements.

Note

- concurrent engineering : 동시공학
- systematic : 규칙적인, 체계적인
- concurrent design : 동시설계
- manufacture : 제조
- outset : 시초, 발단
- life cycle : 수명
- disposal : 처분, 처리

Quality Function Deployment(QFD)

A system for translating consumer/customer requirements into appropriate company requirements at each stage from research and product development to engineering and manufacturing to marketing/sales and distribution.

Note

- quality function deployment : 품질 기능 전개
- consumer : 소비자
- customer : 고객
- appropriate : 고유한
- development : 개발
- distribution : 분배, 공급

Product Life Cycle

The product life cycle goes through many phases, involves many professional disciplines, and requires many skills, tools and processes. Product life cycle has to do with the life of a product in the market with respect to business/commercial costs and sales measures; whereas product lifecycle management has more to do with managing descriptions and properties of a product through its development and useful life, mainly from a business/engineering point of view.

Note

- product life cycle : 제품 수명
- go through : 통과하다
- discipline : 훈련
- with respect to : ~에 관하여
- whereas : 그런데
- product lifecycle management(PLM) : 제품 수명 관리
- point of view : 관점

Life Cycle Engineering(LCE)

Life Cycle Engineering(LCE), sometimes referred to as life cycle design, is a decision-making methodology which considers environmental, performance and cost requirements throughout the duration of a facility. It focuses on the entire life cycle of a design from conceptualization through manufacture to disposal.

Note

- life cycle engineering(LCE) : 수명 공학
- decision-making : 의사결정의
- methodology : 방법론
- environmental : 환경의
- performance : 고성능의
- duration : 기간
- facility : 설비
- focus on ~ : ~에 초점을 맞추다
- conceptualization : 개념화
- disposal : 처분

Sequential Engineering(SE)

Traditional engineering, also known as sequential engineering, is the process of marketing, engineering design, manufacturing, testing and production where each stage of the development process is carried out separately, and the next stage cannot start until the previous stage is finished. Therefore the information flow is only in one direction, and it is not until the end of the chain that errors, changes and corrections can be relayed to the start of the sequence, causing estimated costs to be under predicted. This can cause many problems; such as time consumption due to many modifications is being made as each stage does not take into account the next. This method is hardly used today, as the concept of concurrent engineering is more efficient.

Note

- traditional : 전통적인
- carry out : 수행하다
- estimate : 평가하다
- take into account : 참작하다
- sequential engineering : 순차공학
- correction : 수정, 교정
- predict : 소비
- concurrent engineering : 동시공학

Technology Life Cycle

Most new technologies follow a similar technology lifecycle describing the technological maturity of a product. This is not similar to a product life cycle, but applies to an entire technology, or a generation of a technology.

Technology adoption is the most common phenomenon driving the evolution of industries along the industry lifecycle. After expanding new uses of resources they end with exhausting the efficiency of those processes, producing gains that are first easier and larger over time then exhaustingly more difficult.

Note

- technology life cycle : 기술 수명
- maturity : 성숙, 완성
- adoption : 채용, 채택
- is not similar to ~ : ~과 유사하지 않다
- phenomenon : 현상
- evolution : 발달
- resources : 자원
- exhaust : 소모하다

Product Management

Product management is an organizational function within a company dealing with the planning or marketing of a product or products at all stages of the product lifecycle. Product Management is also a collective term used to describe the broad sum of diverse activities performed in the interest of delivering a particular product to market.

Note

- product management : 제품 관리
- deal with : ~을 다루다
- diverse : 다양한, 각양각색의

09 Structural Analysis

Structural analysis

Structural analysis comprises the set of physical laws and mathematics required to study and predicts the behavior of structures. The subjects of structural analysis are engineering artifacts whose integrity is judged largely based upon their ability to withstand loads; they commonly include buildings, bridges, aircraft, and ships. Structural analysis incorporates the fields of mechanics and dynamics as well as the many failure theories. From a theoretical perspective the primary goal of structural analysis is the computation of deformations, internal forces, and stresses. In practice, structural analysis can be viewed more abstractly as a method to drive the engineering design process or prove the soundness of a design without a dependence on directly testing it.

Note

- structural analysis : 구조 해석
- comprise : 포함하다, 구성하다
- artifact : 인공물
- integrity : 완전(한 상태)
- withstand : 잘 견디는, 버티는
- incorporate : 혼합하다
- A as well as B : A뿐만 아니라 B도 역시
- perspective : 견해
- deformation : 변형
- abstractly : 추상적으로
- soundness : 건전

Analytical method

To perform an accurate analysis a structural engineer must determine such information as structural loads, geometry, support conditions, and materials properties. The results of such an analysis typically include support reactions, stresses and displacements. This information is then compared to criteria that indicate the conditions of failure. Advanced structural analysis may examine dynamic response, stability and nonlinear behavior.

There are three approaches to the analysis: the mechanics of materials approach (also known as strength of materials), the elasticity theory approach (which is actually a special case of the more general field of continuum mechanics), and the finite element approach.

Note

- reaction : 반응, 작용
- displacement : 변위
- nonlinear behavior : 비선형 거동
- elasticity theory : 탄성 이론
- stress : 응력
- criteria : 기준
- strength : 강도
- finite element : 유한 요소

Elasticity methods

Elasticity methods are available generally for an elastic solid of any shape. Individual members such as beams, columns, shafts, plates and shells may be modeled. The solutions are derived from the equations of linear elasticity. The equations of elasticity are a system of 15 partial differential equations. Due to the nature of the mathematics involved, analytical solutions may only be produced for relatively simple geometries. For complex geometries, a numerical solution method such as the finite element method is necessary.

Many of the developments in the mechanics of materials and elasticity approaches have been expounded or initiated by Stephen Timoshenko.

Note

- elasticity methods : 탄성 방법
- elastic solid : 탄성체
- beam : 보
- column : 기둥
- shaft : 축
- derive : 끌어내다
- partial differential equation : 편미분 방정식
- numerical solution : 수치 해
- expound : 상술하다. 해설하다

Finite Element Method(FEM)

Finite element method models a structure as an assembly of elements or components with various forms of connection between them. Thus, a continuous system such as a plate or shell is modeled as a discrete system with a finite number of elements interconnected at finite number of nodes. The behavior of individual elements is characterized by the element's stiffness or flexibility relation, which altogether leads to the system's stiffness or flexibility relation. To establish the element's stiffness or flexibility relation, we can use the mechanics of materials approach for simple one-dimensional bar elements, and the elasticity approach for more complex two- and three-dimensional elements. The analytical and computational developments are best effected throughout by means of matrix algebra.

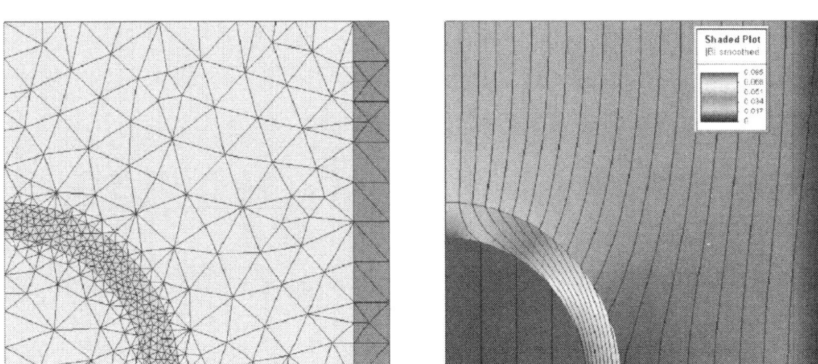

[Figure 9-1] Mesh image and solution

Note

- finite element method : 유한요소법
- finite : 유한의
- behavior : 작용
- flexibility : 유연성
- matrix algebra : 행렬대수
- discrete : 분리된
- interconnect : 상호 연결되다
- stiffness : 강성
- altogether : 전체

Application of FEM

Early applications of matrix methods were for articulated frameworks with truss, beam and column elements; later and more advanced matrix methods, referred to as "finite element analysis", model an entire structure with one-, two-, and three-dimensional elements and can be used for articulated systems together with continuous systems such as a pressure vessel, plates, shells, and three-dimensional solids. Commercial computer software for structural analysis typically uses matrix finite-element analysis, which can be further classified into two main approaches: the displacement or stiffness method and the force or flexibility method. The stiffness method is the most popular by far thanks to its ease of implementation as well as of formulation for advanced applications.

Note

- articulated : 굴절된
- finite element analysis : 유한요소 해석
- pressure vessel : 압력 용기
- classify : 분류하다
- displacement : 변위
- commercial : 상업적인
- stiffness : 강성
- implementation : 이행, 완성

Finite Element Analysis(FEA)

Finite element analysis is a computer simulation technique used in engineering analysis. It uses a numerical technique called the finite element method(FEM). There are many finite element software packages, both free and proprietary. Development of the finite element method in structural mechanics is usually based on an energy principle such as the virtual work principle or the minimum total potential energy principle.

In general, there are three phases in any computer-aided engineering task:

- **Pre-processing** : defining the finite element model and environmental factors to be applied to it.
- **Analysis solver** : solution of finite element model.
- **Post-processing** of results using visualization tools.

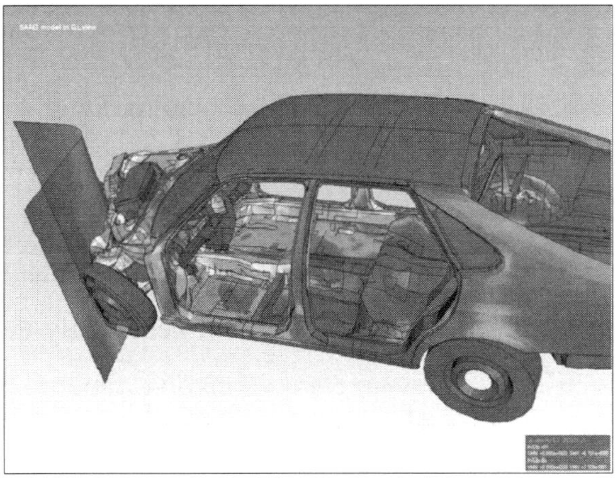

[Figure 9-1] FEA visualization

Note

- finite element analysis : 유한요소해석
- proprietary : 전매 특허의
- finite element method : 유한요소법
- pre-processing : 전처리
- be applied to ~ : ~에 적용되다
- post-processing : 후처리

10 Machine Design

Safety Factor

A safety factor was originally a number by which the ultimate tensile strength of a material was divided in order to obtain a value of "working stress" or "design stress." These design stresses, in turn, were often used in highly simplified calculations that made no allowance for such factors as stress concentration, impact, fatigue, differences between properties of the material in the standard test specimen and in the manufactured part, and so on. Modern engineering design give a rational accounting for all factors possible, leaving relatively few items of uncertainty to be covered by a safety factor, which is commonly in the range of 1.25 to 4

Modern engineering practice also bases the safety factor on the significant strength of the materia-not necessarily the static tensile strength. If the significant stresses involve fatigue, then the safety factor is based on the fatigue strength; if brittle fracture is the expected mode of failure, then the factor is based on tensile strength, and so forth. Thus the safety factor can be defined as

$$\text{safety factor} = \frac{\text{signifcant strength of the material}}{\text{corresponding signinicant stress, form normal loads}}$$

Note

- fact of safety(safety factor) : 안전계수
- ultimate tensile strength : 극한인장강도

- working stress : 작동응력
- make no allowance for ~ : ~를 고려하지 않다
- stress concentration : 응력집중
- impact : 충격
- fatigue : (재료의)피로
- test specimen : 시험편
- static tensile : 정적 인장응력
- fatigue strength : 피로 강도
- brittle fracture : 취서파괴
- not necessarily : 반드시(~은 아니다)
- and so forth : 기타 등등

Fracture mechanics

Very briefly, the fracture mechanics approach begin with the assumption that all real material contain cracks of some size-even if only submicroscopic. If brittle fracture occurs, it is because the conditions of loading and environment(primarily temperature) are such that they cause an almost instantaneous propagation to failure of one or more of the original crack. If there is fatigue loading, the initial cracks may grow very slowly until one of them reaches a critical size(for loading, geometry, and environment involved), at which time total fracture occurs.

Note

- fracture mechanics : 파괴역학
- submicroscopic : 극 미소 물체의
- crack : 균열, 틈
- brittle fracture : 취성 파괴
- instantaneous propagation : 순간적인 확대
- fatigue loading : 피로하중
- critical size : 임계 크기

Reliability

A concept closely related to safety factor is reliability. If 100 "identical" parts are put into service and two fail, then the parts proved to be 98 percent reliable(which might or might not be good enough). Although the reliability concept finds considerably more application with parts subjected to wear and fatigue loadings, we introduce it here in the simpler context of static loading. The usefulness of the reliability approach depends on having adequate information on the statistical distribution of loading applied to parts in service, form which can be calculated the significant stress, and the significant strength of production runs of manufactured parts.

Note

- reliability : 신뢰성
- subject to ~ : ~를 받는
- statistical distribution : 통계 분포
- identical : 똑같은
- context : 정황, 배경
- in service : 작동 중인

S-N curve

In high-cycle fatigue situations, materials performance is commonly characterised by an S-N curve, also known as a Wöhler curve. This is a graph of the magnitude of a cyclical stress (S) against the logarithmic scale of cycles to failure (N). S-N curves are derived from tests on samples of the material to be characterised where a regular sinusoidal stress is applied by a testing machine which also counts the number of cycles to failure. This process is sometimes known as coupon testing. Each coupon test generates a point on the plot though in some cases there is a run out where the time to failure exceeds that available for the test. Analysis of fatigue data requires techniques from statistics, especially survival analysis and linear regression.

Note

- fatigue : 피론
- be characterized : 특징이 부여되다
- logarithmic scale : 로그 스케일
- performance : 성능
- cyclical stress : 반복 응력
- sinusoidal stress : 정현(正弦)응력

Corrosion

Corrosion is the degradation of a material(normally a metal) by chemical or electrochemical reaction with its environment. Most corrosion result from electrochemical, or galvanic, action. This is a complex phenomenon, giving rise to the specialized discipline of corrosion engineering.

As soon as the slightest crack is formed, the corrosive liquid penetrates it and works on its sharp bottom edge. The combination of the corrosive action and alternating stress at the bottom of the crack is vicious and reduces the safe stress very materially.

Note

- corrosion : 부식
- degradation : 분해
- electrochemical reaction : 전기화학적 반응
- penetrate : 스며들다
- alternating stress : 변동 응력
- crack : 균열

Abrasive wear

The term, "wear" most often refers to abrasive wear, which is due to the rubbing of abrasive particles on a surface. These particles are typically small and hard and have sharp edge-like grains of sand or particles of metal or metal oxide that rub off a wearing metal surface. Common examples include the wearing down of wood or metal with sand or emery paper or with a grinding wheel, the wearing of shoe leather by scuffing against cement sidewalks, the wearing of a plow blade or earth auger during use, and the removal of metal from the journal surface of a rotating shaft by foreign abrasive particles in the lubricant.

Note

- abrasive : 마모의, 연마의
- rubbing : 연마
- emery : 연마사(金剛砂]
- lubricant : 윤활제
  cf. machinery :기계류(집합명사), machine : 기계(보통명사)
- wear : 마모(磨耗), 마멸(磨滅)
- sharp edge : 날카로운 날
- scuff : 발을 질질끌며 걷다

Lubrication

Lubrication is concerned with the reduction of friction, wear, and heating of machine parts which move relative to each other. A lubricant is any substance which, when inserted between the moving surfaces, accomplishes these purpose. In a sleeve bearing a shaft, or journal, rotates or oscillates within a sleeve, or bearing, and the relative motion is sliding in an anti-friction bearing, the main relative motion is rolling. Gear teeth mate with each other by combination of rolling and sliding. Piston slide within their cylinders. All these applications require lubrication to reduce friction, and bearing.

Note

- lubrication : 윤활
- friction : 마찰
- sleeve bearing : 슬리브 베어링
- main relative motion : 주 상대운동
- lubricant : 윤활제
- wear : 마모
- reduction gear : 감속 기어
- combination : 조합

Rivet

A rivet is a mechanical fastener. Before it is installed it consists of a smooth cylindrical shaft with a head on one end. The end opposite the head is called the buck-tail. On installation the rivet is placed in a pre-drilled hole. Then the tail is “upset” (i.e. deformed) so that it expands to about 1.5 times the original shaft diameter and holds the rivet in place. To distinguish between the two ends of the rivet, the original head is called the factory head and the deformed end is called the shop head or buck-tail.

Note

- fastener : 파스너, 조임구
- installation : 장착, 설치
- cylindrical shaft : 원통 축
- distinguish : 구별하다

Lap joint

Lap joint is meant an arrangement which consists of overlapping plates held together by one or more rows of rivets. If one row of rivets is used, the arrangement is called a single-riveted lap joint, and with two rows are used, it is called a double-riveted lap joint.

Note

- lap joint : 겹치기 이음
- overlapping : 겹치기
- consist of : ~으로 구성된
- row : 열

Butt joint

When plates butt against each other and are joined by overlapping plates or strips, the connection is called a butt joint.

Note

- butt joint : 맞대기 이음
- strips : 대판

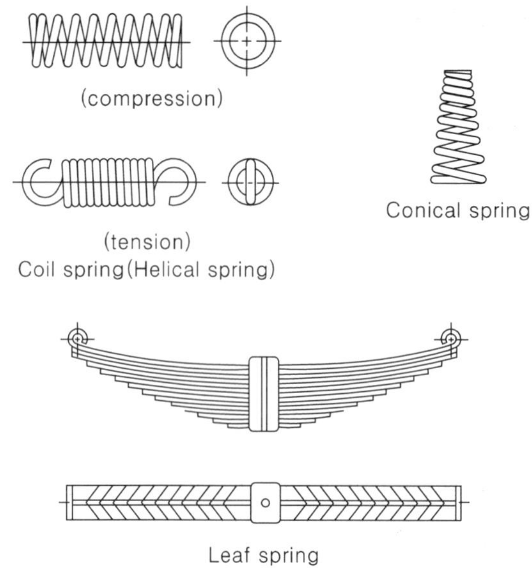

[Figure 10-1] Spring

Spring

Springs are made in a variety of forms, depending upon the class of service for which they are intended. Among the common forms used to a considerable extent in connection with machinery, are as follows

(a) Helical spring
(b) Spiral spring
(c) Conical spring
(d) Leaf spring

Note

- spring : 용수철, 스프링
- variety of forms : 여러 가지 형태
- helical spring : 코일 스프링
- spiral spring : 나선형 스프링
- conical spring : 원뿔 스프링
- leaf spring : 판 스프링

Helical spring

A Coil spring, also known as a helical spring, is a mechanical device, which is typically used to store energy and subsequently release it, to absorb shock, or to maintain a force between contacting surfaces. They are made of an elastic material formed into the shape of a helix which returns to its natural length when unloaded.

Coil springs are a special type of torsion spring, the material of the spring acts in torsion when the spring is compressed or extended.

Note

- helical spring : 코일 스프링
- device : 장치
- store : 저장하다
- absorb shock : 충격흡수
- helix : 나선
- torsion spring : 비틀림 스프링
- elastic material : 탄성재료

Leaf spring

Originally called laminated or carriage spring, a leaf spring is a simple form of spring, commonly used for the suspension in wheeled vehicles. It is also one of the oldest forms of springing.

Sometimes referred to as a semi-elliptical spring or cart spring, it takes the form of a slender arc-shaped length of spring steel of rectangular cross-section. The center of the arc provides location for the axle, while tie holes are provided at either end for attaching to the vehicle body.

Note

- leaf spring : 판 스프링
- laminated : (얇은)판으로 된
- carriage spring : 차량용 스프링
- rectangular : 직사각형
- cross-section : 단면
- vehicle body : 차체(車體)

Conical spring

Conical springs are generally used to resist a compression and are made of round or rectangular stock. They are applicable where the space is limited, and where there is no necessity for great deflection. For conical spring made of round stock and loaded, the shearing stress in the material is as follows ;

$$S_s = \frac{8QD_2}{\pi d^3}$$

Note

- conical spring : 원뿔형 스프링
- round stock : 원형재
- rectangular stock : 각재
- rectangular deflection : 축방향 처짐

Belt

Ordinarily, belts are used to transmit power between two parallel shaft. The shafts must be separated a certain minimum distance, which is dependent upon the type of belt used, in order to work most efficiently. Belts have the following characteristics.

1) They may be used for long center distance.
2) Because of the slip and creep of belts, the angular velocity ratio between the two shafts is neither constant nor exactly equal to the ratio of the pulley diameters.
3) By employing step pulleys, economical means of changing the velocity ratio may be obtained.
4) When using the flat belts, clutch action may be obtained by shifting the belt from a loose pulley to a tight pulley

Note

- flexible machine element : 유연성있는 기계요소
- parallel shaft : 평행축
- slip : 미끄럼
- angular velocity ratio : 각속도비
- step pulley : 단차(段車)
- economical mean : 경제적 수단
- tight pulley : 고정 풀리

Gear

Whenever rotary motion must be transferred from one shaft to another, gear should be considered as one of the possible means. They have the advantage of a constant velocity ratio.

Modern demands for greater loads and higher speeds, with the conflicting requirement of weight reduction, require that the machine designer be fully informed on current design practices. This must include a knowledge not only of materials, fatigue and manufacturing methods as applied to gear but also of the accepted methods of analysis and design.

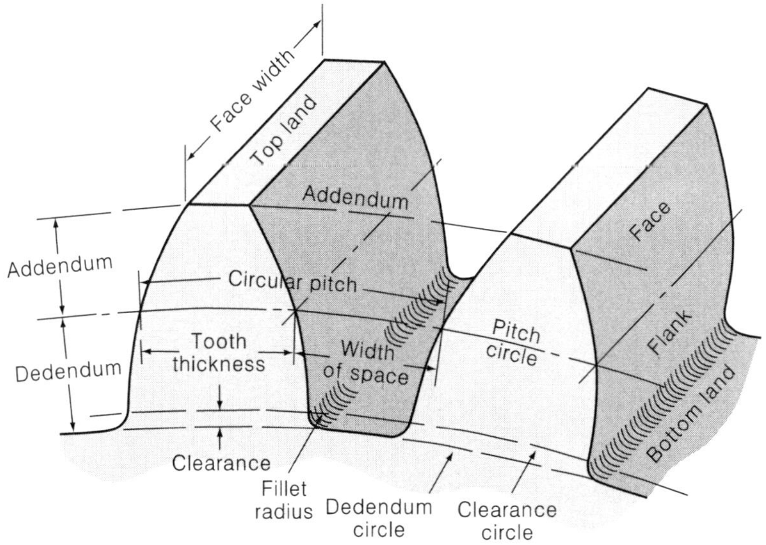

[Figure 10-2] Gear

Note

- rotary motion : 회전운동
- constant velocity ratio : 일정한 속도비
- advantage : 이점
- conflicting : 모순되는, 상반되는
- not only ~ but also : ~뿐만 아니라 ~ 또한
- fatigue : 피로
- addendum : 이끝높이
- dedendum : 이 뿌리 높이
- top land : 이끝면
- addendum circle : 이 봉우리(끝)원
- dedendum circle : 이 골(뿌리)원
- circle pitch : 원주피치
- pitch circle : 피치원
- clearance : 크리어런스
- flank : 치면
- tooth thickness : 이 두께

11 Vibration

Importance of the study of vibration

Early scholars in the field of vibration concentrated their efforts on understanding the natural phenomena and developing mathematical theories to describe the vibration of physical system. In recent times, many investigations have been motivated by the engineering applications of vibration such as the design of machines, foundations, structures, engines, turbines, and control system.

In turbines, vibrations cause spectacular mechanical failures. Naturally, the structures designed to support heavy centrifugal machines, like motors and turbines, or reciprocating machines, like steam and gas engines and reciprocating pumps, are also subjected to vibration. In all these situations, the structure or machine component subjected to vibration can fail because of material fatigue resulting from the cyclic variation of the induced stress.

Note

- vibration : 진동
- undergo : ~를 받다
- breathing : 호흡
- oscillatory motion : 진동운동
- larynges : 후두(喉頭)
- investigation : 조사, 연구
- centrifugal machine : 원심기계
- cyclic vibration of the induced stress : 유발된 응력의 주기적인 변화

Elementary part of vibrating system

A vibratory system, in general, includes a means for storing potential energy(spring or elasticity), a means for storing kinetic energy(mass or inertia), and a mean by which energy is gradually lost(damper).

The vibration of system involves the transfer of its potential energy to kinetic energy and kinetic energy to potential energy, alternately. If system damped, some energy is dissipated in each cycle of vibration and must be replaced by an external source if a state of steady vibration is to be maintained.

Note

- vibratory : 진동을 일으키는
- store : 저장하다
- potential energy : 위치 에너지, 잠재에너지
- elasticity : 탄성
- kinetic energy: 운동에너지
- in each : 관성
- alternately : 양자택일론
- damp : (진폭을)줄이다.
- dissipate : 없어지다,

Periodic motion of vibration

A vibration in its general sense is a periodic motion, i.e. a motion which repeats itself in all its particulars after a certain interval of time, called the period of the vibration. A plot of displacement against the time may be a curve of considerable complication.

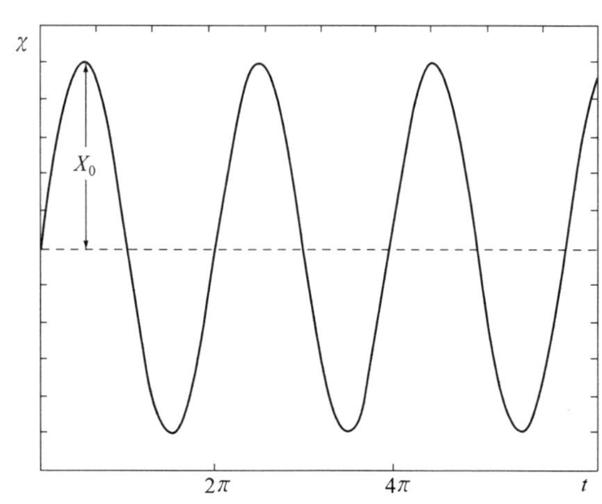

[Figure 11-1] Motion curve

As an example, Figure 11-1 shows the motion curve observed on the bearing pedestal of a steam turbine. The simplest kind of periodic motion is a harmonic motion; in it the relation between x and t may be expressed by

$$x = X_0 \sin \omega t$$

The maximum value of displacement is X_0, called the amplitude of the vibration.

The period T usually is measured in seconds; its reciprocal is frequency of vibration($1/T = f$), measured in cycle per second.

Note

- periodic motion : 주기적 운동
- interval of time : 시간간격
- pedestal : 축받이
- amplitude : 진폭
- frequency : 진동수, 주파수
- particulars : 상세, 특수
- motion curve : 운동곡선
- harmonic motion : 조화운동
- reciprocal : 역수

Free and forced vibration

Free vibration. If a system, after an initial disturbance, is left to vibrate on its own, the ensuing vibration is known as free vibration. No external force acts on the system. The oscillation of a simple pendulum is an example of free vibration.

Forced vibration. If a system is subjected to an external force(often, a repeating type of force), the resulting vibration is known as forced vibration. The oscillation that arises in machines such as diesel engines is an example of force vibration.

If the frequency of the external force coincides with on of the natural frequencies of the system, a condition known as resonance occurs, and the system undergoes dangerously large oscillations.

Failure of such structures as building, bridges, turbines, and airplane wings have been associated with the occurrence of resonance.

Note

- free vibration : 자유진동
- forced vibration : 강제 진동
- disturbance : 교란
- ensuing : 뒤이은
- external force : 외력
- pendulum : 진자
- coincide : 동시에 일어나다
- resonance : 공진
- undergo : 만나다

Undamped and damped vibration

If no energy is lost or dissipated in friction or other resistance during oscillation, the vibration is known as undamped vibration. If any energy is lost in this way, however, it is called damped vibration. In many physical systems, the amount of damping is so small that it can be disregarded for most engineering purposes. However, consideration of damping becomes extremely important in analyzing vibratory systems near resonance.

Note

- undamped vibration : 비감쇠 진동
- damped vibration : 감쇠진동
- dissipate : 없어지다
- friction : 마찰
- oscillation : 진동
- disregarded : 무시하다
- consideration : 연구, 고려
- resonance : 공명

Spring elements

Linear spring is a type of mechanical link that is generally assumed to have negligible mass and damping. A force is developed in the spring whenever there is relative motion between the two ends of the spring. The spring force is proportional to the amount of deformation and is given by

$$F = kx$$

Where F is spring constant, x is the deformation(displacement of one end with respect to the other), and k is the spring stiffness or spring constant. If we plot a graph between F and x, the result is straight line according to equation.

Note

- linear spring : 선형 스프링
- mass : 질량
- relative motion : 상대운동
- spring constant : 스프링 상수
- displacement : 변위
- negligible : 무시할 수 있는
- damping : 감쇠
- proportional : 비례하는
- deformation : 변형
- spring stiffness : 스프링 강성

Damping elements

In many practical systems, the vibrational energy is gradually converted to heat or sound. Due to the reduction in the energy, the response, such as the displacement of the system, gradually decreases. The mechanism by which the vibrational energy is gradually converted into heat or sound is known as damping. Although the amount of energy converted into heat or sound is relatively small, the consideration of damping becomes important for an accurate prediction of the vibration response of a system. A damper is assumed to have neither mass nor elasticity, and damping force exists only if there is relative velocity between the two ends of the damper.

Note

- damping element : 감쇠요소
- convert : 변환시키다
- response : 응답
- gradually : 점점
- reduction : 축소
- displacement : 변위

- decrease : 감소하다
- elasticity : 탄성
- accurate prediction : 정확한 예측

cf. neither A nor B : A도 아니고 B도 아닌

Neither he nor his wife has arrived. : 그도, 그의 부인도 도착하지 않았다.

Degree of freedom

A mechanical system is said to have one degree of freedom if its geometrical position can be expressed at any instant by one number only. Generally if it takes n numbers to specify the position of a mechanical system, it is said to have n degrees of freedom.

Note

- degree of freedom : 자유도
- geometrical position : 기하학적 위치
- at any instant : 어떤 순간에
- specify : 지시하다, 규정하다

Natural frequency

If a system, after an initial disturbance, is left to vibrate on its own, the frequency with which it oscillates without external forces is known as it natural frequency. As will be seen later, a vibratory system having n degrees of freedom will have, in general, n distinct natural frequencies of vibration. Equation of natural frequency(in rad/s) is as follows ;

$$\omega_n = \sqrt{\frac{k}{m}}$$

Where m is mass of the system

k is spring constant of the system

Note

- natural frequency : 고유 진동수
- frequency : 주파수
- external force : 외력
- distinct : 별갱의
- initial disturbance : 초기교란
- oscillate : 진동하다
- degrees of freedom : 자유도

Vibration absorbers

A machine or system may experience excessive vibration if it is acted on by a force whose excitation frequency nearly coincides with a natural frequency of the machine of system. In such cases, the vibration of the machine or system can be reduced by using a vibration neutralizer of dynamic vibration absorber, which is simply another spring mass system. The dynamic vibration absorber is designed such that the natural frequencies of the resulting system are away from the excitation frequency. We shall consider the analysis of a dynamic vibration absorber by idealizing the machine as a single degree of freedom system.

Note

- vibration absorber : 진동 흡진기
- coincide : 동시에 일어나다
- natural frequency : 고유 진동수
- single degree or freedom system : 1자유도 계
- neutralizer : 중화기(中和機)
- excessive : 과도의
- excitation frequency : 가진 주파수

12 Manufacturing

Casting

Casting is a manufacturing process by which a liquid material is (usually) poured into a mold, which contains a hollow cavity of the desired shape, and then allowed to solidify. The solid casting is then ejected or broken out to complete the process.

Casting may be used to form hot liquid metals or various materials that cold set after mixing of components (such as epoxies, concrete, plaster and clay). Casting is most often used for making complex shapes that would be otherwise difficult or uneconomical to make by other methods.

Note

- casting : 주조
- liquid material : 유동성 재료
- mold : 주형
- cavity : 공동(空洞)
- solidify : 응고시키다
- manufacture process : 제조공정
- pour : 붓다
- hollow : 속이 빈
- plaster : 석고반죽
- clay : 점토

Die casting

Die casting is the process of forcing molten metal under high pressure into mold cavities (which are machined into dies). Most die castings are made from nonferrous metals, specifically zinc, copper, and aluminum based alloys, but ferrous metal die castings are possible. The die casting method is especially suited for applications where many small to medium sized parts are needed with good detail, a fine surface quality and dimensional consistency.

Note

- die casting : 다이캐스팅, 압력주조
- cavity : 공동
- nonferrous metal : 비철금속
- ferrous : 철을 함유한
- surface quality : 표면질
- consistency : 일관성

Shrinkage

The shrinkage caused by solidification can leave cavities in a casting, weakening it. Risers provide additional material to the casting as it solidifies. The riser (sometimes called a "feeder") is designed to solidify later than the part of the casting to which it is attached. Thus the liquid metal in the riser will flow into the solidifying casting and feed it until the casting is completely solid. In the riser itself there will be a cavity showing where the metal was fed. Risers add cost because some of their material must be removed, by cutting away from the casting which will be shipped to the customer.

Note

- shrinkage : 수축
- solidification : 응고
- casting : 주조
- riser : 라이저
- cavity : 기공

Welding

Welding is a fabrication process that joins materials, usually metals or thermoplastics, by causing coalescence. This is often done by melting the workpieces and adding a filler material to form a pool of molten material (the weld puddle) that cools to become a strong joint, with pressure sometimes used in conjunction with heat, or by itself, to produce the weld. This is in contrast with soldering and brazing, which involve melting a lower-melting-point material between the workpieces to form a bond between them, without melting the workpieces.

Note

- welding : 용접
- thermoplastics : 열가소성 물질
- workpiece : 작업물
- puddle : 웅덩이
- in contrast with ~ : ~와 대비되는
- brazing : 납땜하다
- fabrication process : 제조공정
- coalescence : 융합
- filler material : 충진 물질
- conjunction : 결합
- soldering : 납땜

Arc welding

Arc welding uses a welding power supply to create an electric arc between an electrode and the base material to melt the metals at the welding point. They can use either direct (DC) or alternating (AC) current, and consumable or non-consumable electrodes. The welding region is sometimes protected by some type of inert or semi-inert gas, known as a shielding gas, and/or an evaporating filler material. The process of arc welding is widely used because of its low capital and running costs.

Note

- arc welding : 아크 용접
- power supply : 전원공급장치
- alternating current : 교류 전류
- shielding gas : 차폐 기체
- running cost : 가공비, 운용비
- electrode : 전극봉
- direct current : 직류 전류
- consumable electrode : 소모 전극
- inert gas : 불활성 기체

Gas welding

The most common gas welding process is oxyfuel welding, also known as oxyacetylene welding. It is one of the oldest and most versatile welding processes, but in recent years it has become less popular in industrial applications. It is still widely used for welding pipes and tubes, as well as repair work. It is also frequently well-suited, and favored, for fabricating some types of metal-based artwork. Oxyfuel equipment is versatile, lending itself not only to some sorts of iron or steel welding but also to brazing, braze-welding, metal heating (for bending and forming), and also oxyfuel cutting.

Note

- gas welding : 가스 용접
- oxyacetylene : 산소-아세틸렌
- versatile : 다용도의
- repair work : 수리작업
- favor : 지지하다.
- fabricating : 제조
- oxyfuel equipment : 산소용접장비
- brazing : 납땜
- oxyfuel cutting : 산소절단

Forging

Forging is the term for shaping metal by using localized compressive forces. Cold forging is done at room temperature or near room temperature. Hot forging is done at a high temperature, which makes metal easier to shape and less likely to fracture. Warm forging is done at intermediate temperature between room temperature and hot forging temperatures. Forged parts can range in weight from less than a kilogram to 170 metric tons. Forged parts usually require further processing to achieve a finished part.

Note

- forging : 단조(鍛造)
- cold forging : 냉간 단조
- hot forging : 열간 단조
- warm forging : 온간단조
- compressive : 압축력이 있는
- room temperature : 실내온도
- fracture : 파손

Rolling

Rolling is a fabricating process in which the metal, plastic, paper, glass, etc. is passed through a pair (or pairs) of rolls. There are two types of rolling process, flat and profile rolling. In flat rolling the final shape of the product is either classed as sheet (typically thickness less than 3 mm, also called "strip") or plate (typically thickness more than 3 mm). In profile rolling, the final product may be a round rod or other shaped bar such as a structural section (beam, channel, joist etc).

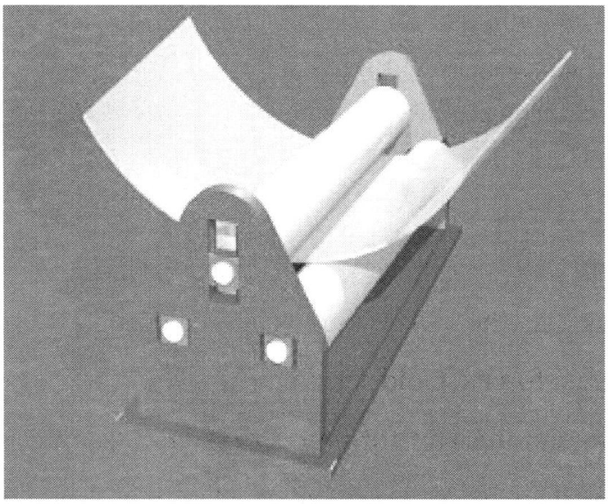

[Figure 12-1] Rolling

Note

- rolling : 압연
- a pair of : 한 쌍의
- profiled rolling : 형재 압연
- fabricating : 제조
- flat rolling : 판재 압연
- joist : 들보

Drawing

Drawing is a manufacturing process for producing a wire, bar or tube by pulling on a material until it increases in length. This is typically accompanied by a thinning out of the material, usually through a reduction in the girth.

This technique has long been used to produce flexible metal wire by drawing the material through a series of dies (holes) of decreasing size. These dies are manufactured from a number of materials for wire drawing, the most common being Tungsten Carbide and Diamond. Diamond dies can either be synthetic or natural, synthetic diamond is usually used in the early stages of the drawing process, whereas natural diamond dies are used in the final stages. More recently drawing has been used with molten glass to produce high quality optical fibers.

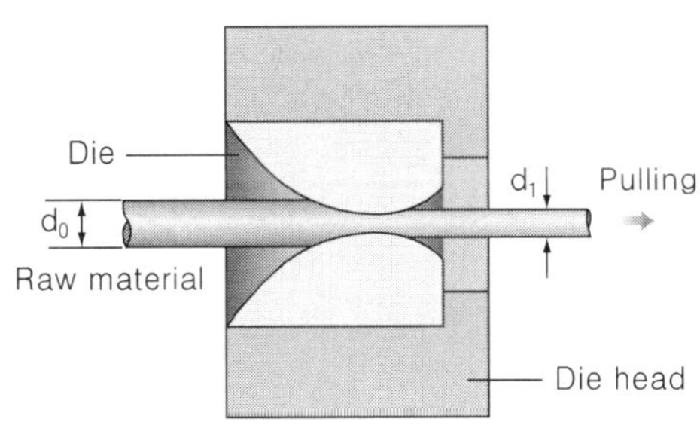

[Figure 12-2] Drawing

Note

- drawing : 인발
- bar : 봉
- tube : 관
- girth : 둘레
- flexible metal : 가요성(可撓性) 금속
- a series of : 일련의
- synthetic diamond : 인조 다이아몬드
- optical fiber : 광섬유

Extrusion

Extrusion is a manufacturing process used to create long objects of a fixed cross-sectional profile. A material, often in the form of a billet, is pushed and/or drawn through a die of the desired profile shape. Hollow sections are usually extruded by placing a pin or piercing mandrel inside of the die, and in some cases positive pressure is applied to the internal cavities through the pin. Extrusion may be continuous (producing indefinitely long material) or semi-continuous (producing many short pieces). Some materials are hot drawn while others may be cold drawn.

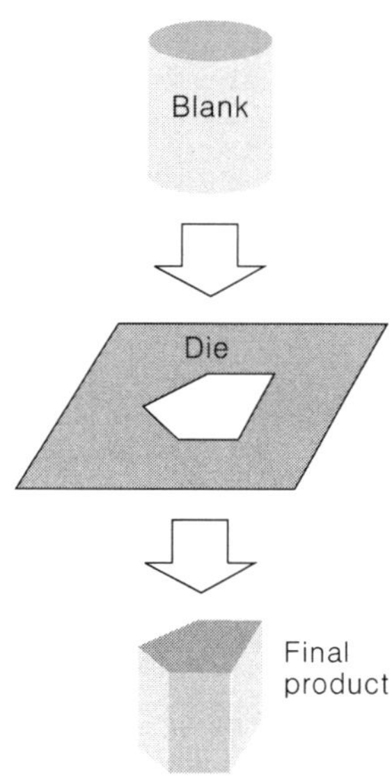

[Figure 12-3] Extrusion

Note

- extrusion : 압출
- cross-sectional profile : 단면형상
- billet : 강편(鋼片)
- hollow section : 중공단면
- piercing : 천공(穿孔)
- mandrel : 맨드릴, 심봉(心棒)

Heat treatment

Heat treatment is a method used to alter the physical, and sometimes chemical, properties of a material. The most common application is metallurgical. Heat treatments are also used in the manufacture of many other materials, such as glass. Heat treatment involves the use of heating or chilling, normally to extreme temperatures, to achieve a desired result such as hardening or softening of a material. Heat treatment techniques include annealing, case hardening, precipitation strengthening, tempering and quenching.

Note

- heat treatment : 열처리
- alter : 바꾸다
- metallurgical : 야금학적인
- chilling : 냉각
- softening : 연화
- case hardening : 표면경화
- tempering : 불림
- quenching : 담금질
- precipitation strengthening : 석출강화(析出 强化)

Annealing

Annealing is a technique used to recover cold work and relax stresses within a metal. Annealing typically results in a soft, ductile metal. When an annealed part is allowed to cool in the furnace, it is called a "full anneal" heat treatment. When an annealed part is removed from the furnace and allowed to cool in air, it is called a "normalizing" heat treatment. During annealing, small grains recrystallize to form larger grains. In precipitation hardening alloys, precipitates dissolve into the matrix, "solutionizing" the alloy.

Note

- annealing : 풀림
- relax : 긴장을 늦추다
- furnace : 노
- grain : 결정
- hardening : 용해하다
- normalizing : 불림
- cold work : 냉간 가공
- ductile metal : 연성금속
- heat treatment : 열처리
- precipitation : 침전
- solutionizing : 용체화처리
- recrystallize : 재결정화되다

Quenching and tempering

To harden by quenching, a metal (usually steel or cast iron) must be heated into the austenitic crystal phase and then quickly cooled. Depending on the alloy and other considerations (such as concern for maximum hardness vs. cracking and distortion), cooling may be done with forced air or other gas (such as nitrogen), oil, polymer dissolved in water, or brine. Upon being rapidly cooled, a portion of austenite (dependent on alloy composition) will transform to martensite, a hard brittle crystalline structure.

Untempered martensite, while very hard and strong, is too brittle to be useful for most applications. A method for alleviating this problem is called tempering.

Note

- harden : 경화시키다
- cast iron : 주철
- cracking : 열분해
- nitrogen : 질소
- brittle : 부서지기 쉬운
- quenching : 담금질
- hardness : 경화
- distortion : 비틀림
- brine : 소금물
- alleviating : 완화시키는

cf. too ~ to ~ : 매우 ~해서, ~할 수 없다.

ex) A baby is too young to walk. : 아기는 너무 어려서 걸을 수 없다

Machining

Conventional machining, one of the most important material removal process, is a collection of material-working processes in which power-driven machine tools, such as lathes, milling machines, and drill presses are used with a sharp cutting tool to mechanically cut the material to achieve the desired geometry. Machining is a part of the manufacture of almost all metal products. It is not uncommon for other materials to be machined. A person who specializes in machining is called a machinist. Machining is also a hobby. A room, building, or company where machining is done is called a machine shop.

Note

- machining : 기계가공
- removal : 제거
- lathe : 선반
- milling machine : 밀링 기계
- geometry : 형상
- machinist : 기계제작공
- machine shop : 기계 공장

Built up edge

Because shear is strongest at the contact surface with the cutting tool, the first layer of metal impacting and seizing on it work-hardens more than the rest of the volume of metal. As a consequence of this work hardening, this first layer of metal is stronger than the adjacent metal moving away from the workpiece. Effectively, said first layer becomes part of the tool.

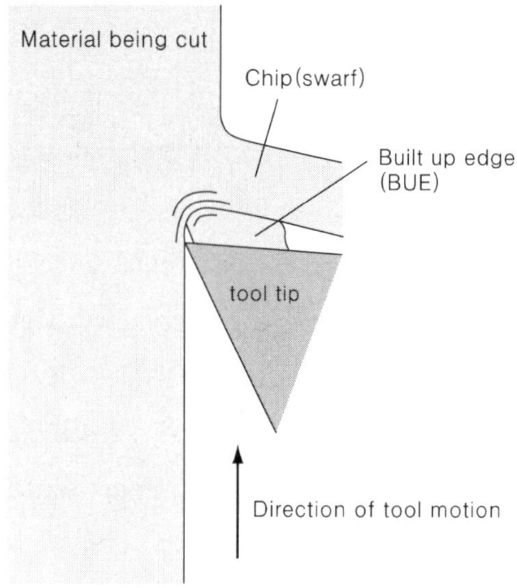

[Figure 12-4] Built up edge

Note

- built up edge : 구성인선(構成刃先)
- shear : 전단
- contact surface : 접촉면
- adjacent : 인접한

Lapping

Lapping is an operation that can be done by hand or machine where two surfaces are rubbed together with an abrasive in the middle to try to get the surfaces as flat and as smooth as possible.

This can take two forms. The first type of lapping (traditionally called grinding), typically involves rubbing a brittle material such as glass against a surface such as iron or glass itself (also known as the "lap" or grinding tool) with an abrasive such as aluminum oxide, emery, silicon carbide, diamond, etc., in between them. This produces microscopic conchoidal fractures as the abrasive rolls about between the two surfaces and removes material from both.

Note

- lapping : 래핑
- abrasive : 연마용의
- aluminum oxide : 알루미늄 산화물
- fracture : 부서지다
- conchoidal : 패각상(貝殼狀)의
- rub : 문지르다
- brittle : 부서지기 쉬운
- emery : (연마용) 금강사
- silicon carbide : 탄화 규소

Polishing

Polishing is the process of creating a smooth and shiny surface by using rubbing or a chemical action.

In metallography polishing is used to create a flat, defect free surface for examination of a metal's microstructure under a microscope. Silicon based polishing pads or diamond solution can be used in the polishing process.

Note

- polishing : 연마
- rub : 마찰하다
- defect free : 결점이 없는
- shiny surface : 윤기나는 표면
- metallography : 금속학
- microstructure : 미세구조

Milling machine

A milling machine is a machine tool used for the shaping of metal and other solid materials. Its basic form is that of a rotating cutter which rotates about the spindle axis (similar to a drill), and a table to which the workpiece is affixed. The cutter and workpiece move relative to each other, generating a toolpath along which material is removed. The movement is precisely controlled, usually with slides and leadscrews or analogous technology. Often the movement is achieved by moving the table while the cutter rotates in one place, but regardless of how the parts of the machine slide, the result that matters is the relative motion between cutter and workpiece. Milling machines may be operated manually or by CNC (computer numerical control).

Note

- milling machine : 밀링 기계
- spindle axis : 스핀들축
- tool path : 공구 경로
- numerical control : 수치제어
- rotating cutting : 회전 절삭공구
- affix : 붙이다
- regardless of ~ : ~와 관계없이

Introduction to lathe

A lathe is a machine tool which spins a block of material to perform various operations such as cutting, sanding, knurling, drilling, or deformation such as metal spinning with tools that are applied to the workpiece to create an object which has symmetry about an axis of rotation.

Lathes are used in woodturning, metalworking, metal spinning, and glassworking. Lathes can be used to shape pottery, the best-known design being the potter's wheel. Most suitably equipped metalworking lathes can also be used to produce most solids of revolution, plane surfaces and screw threads or helices. Ornamental lathes can produce three-dimensional solids of incredible complexity. The material is held in place by either one or two centers, at least one of which can be moved horizontally to accommodate varying material lengths.

Note

- lathe : 선반
- knurling : 널링
- symmetry : 대칭
- potter : 도공
- revolution : 회전
- screw thread : 나사산
- ornamental : 장식용의
- of incredible complexity : 상당하게 복합한
- at least : 최소한
- accommodate : 조절시키다

Parts of lathe

A lathe may or may not have a stand (or legs), which sits on the floor and elevates the lathe bed to a working height. Some lathes are small and sit on a workbench or table, and do not have a stand.

All lathes have a "bed", which is (almost always) a horizontal beam (although some CNC lathes have a vertical beam for bed to ensure that swarf, or chips, falls free of the bed).

Note

- lathe : 선반
- workbench : 작업대
- horizontal : 수평의
- beam : 보
- vertical : 수직의
- swarf : 잘라낸 토막

• Headstock

At one end of the bed (almost always the left, as the operator faces the lathe) is a "headstock". The headstock contains high-precision spinning bearings.

• Spindle

Rotating within the bearings is a horizontal axle, with an axis parallel to the bed, called the "spindle". Spindles are often hollow, and have exterior threads and/or an interior Morse taper on the "inboard" (i.e., facing to the right / towards the bed) by which accessories which hold the workpiece may be mounted to the spindle.

Note

- spindle : 스핀들, 축
- hollow : 중공의
- inboard : 나사산
- axle : 차축
- exterior : 외부의
- interior : 내부의

• Tailstock

The counterpoint to the headstock is the tailstock, sometimes referred to as the loose head, as it can be positioned at any convenient point on the bed, by undoing a locking nut, sliding it to the required area, and then relocking it. The tailstock contains a barrel which does not rotate, but can slide in and out parallel to the axis of the bed, and directly in line with the headstock spindle.

Note

- tailstock : 심압대
- headstock : 주축대
- spindle : 스핀들
- counter point : 대조적인 요소
- convenient : 편리한

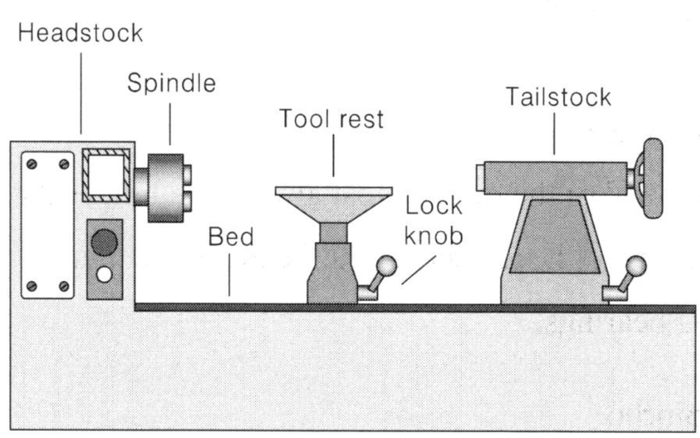

[Figure 12-5] Lathe

13 Mechatronics

Definition of mechatronics

Consider the modern auto-focus, auto-exposure camera. To use the camera all you need to do is point it at the subject and press the button to take the picture. The camera automatically adjusts the focus so that the subject is in focus and automatically adjusts the aperture and shutter speed so that the correct exposure is given.

Such control systems generally use microprocessor as controllers and have electrical sensor extracting information from the mechanical inputs and outputs via electrical actuators to mechanical systems. The term “mechantronics” is used for this integration of microprocessor control system, electrical systems and mechanical systems. A mechatronics system is not just a marriage of electrical and mechanical systems and is more than just a control system; it is a complete integration of all of them.

Note

- mechatronics : mechanics+electronics+informatics
- auto-focus : 자동초점
- auto-exposure : 자동노출
- aperture : 조리개
- controller : 제어장치
- electrical actuator : 전기적 구동장치
- integration : 통합

Sensor and transducers

The term “sensor” is used for an element which produces a signal relating to the quantity being measured. Thus in the case of, say, an electrical resistance temperature element, the quantity being measured is temperature and the sensor transforms an input of temperature into a change in resistance. The term "transducer" is often used in place of the term sensor. Transducers are defined as elements that when subject to some physical change experience a related change. Thus sensors are transducer, however, a measurement system may use transducers, in addition to the sensor, in other parts of the system to convert signals in one form to another form.

Note

- term : 용어
- signal : 신호
- resistance temperature element : 저항 온도요소
- transform : 변형시키다
- transducer : 변환기
- subject to ~ : ~을 조건으로 하는
- convert : 전환하다

1. Reed switch

Figure 13-1 shows the basic form of a reed switch. It consists of two magnetic switch contacts sealed in a glass tube. When a magnet is brought close to the switch, the magnetic reeds are attracted to each other and close the switch contacts. It is a non-contact proximity switch.

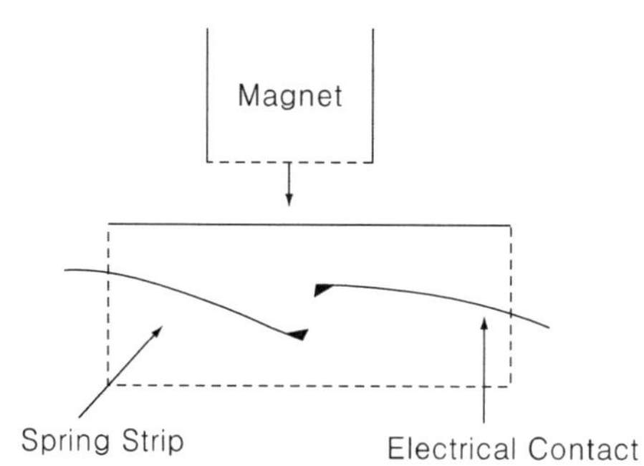

[Figure 13-1] Reed switch

Such a switch is very widely used for checking closure of doors. It is also used with such a devices as tachometers which involve the rotation of a toothed wheel past the reed switch. If one of the teeth has a magnet attached to it, then every time it passes the switch it will momentarily close the contacts and hence produce a current/voltage pulse in the associated electrical circuit

Note

- reed switch : 리드 스위치, 자석에 의하여 작동되는 스위치
- close to : 접근하여
- non-contact proximity switch : 비접촉식 근접 스위치
- tachometer : 회전 속도계
- momentarily : 순간적으로
- hence : 그러므로

2. Photosensitive devices

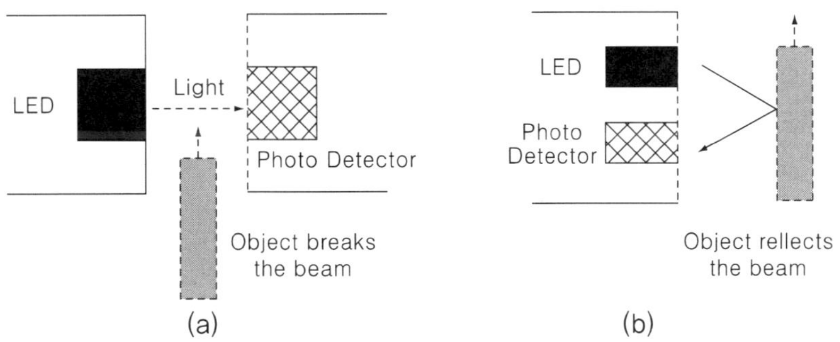

[Figure 13-2] Using photoelectric sensors to detect objects by (a) the object breaking the beam, (b) it reflecting light

Photosensitive devices can be used to detect the presence of an opaque object by it breaking a beam of light, or infrared radiation, falling on such a device of by detecting the light reflected back by the object(figure).

Note

- photosensitive : 광감지형의
- presence : 존재
- infrared radiation : 적외선
- detect : 검출하다
- opaque object : 불투명체

3. Tachogenerator

The tachogenerator is used to measure angular velocity. One form, the variable reluctance tachogenerator, consists of a toothed wheel of ferromagnetic material which is attached to the rotating shaft(figure). A pick-up coil is wound on a permanent magnet. As the rotation, so the teeth move past the coil and the air gap between the coil and the ferromagnetic material changes. The resulting cyclic change in the flux linked produces an alternating e.m.f in the coil.

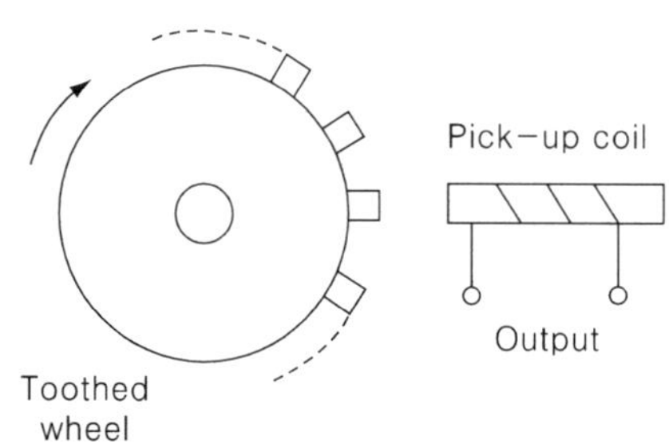

[Figure 13-3] The principle of tachogenerator

Another form of tachogenerator is essentially an A.C. generator. It consists of a coil, termed the rotor, which rotates with the rotating shaft. This coil rotates in the magnetic field produced by a stationary permanent magnet or electromagnetic and so an alternating e.m.f. is induced in it. The amplitude or frequency of this alternating e.m.f. can be used as a measure of the angular velocity of the rotor.

Note

- tachogenerator : 회전속도에 따른 유도기전력을 발생하는 센서의 일종
- angular velocity : 각속도
- reluctance : 자기저항형
- ferromagnetic material : 강자성 재료
- e.m.f : electromotive force, 기전력
- A.C. : Alternating Current, 교류
- magnetic field : 자기장
- induce : 유도하다

4. Strain gauge and load sell

A very commonly used for form of force–measuring transducer is based on the use of electrical resistance stain gauge to monitor the strain produced in some member when stretched, compressed or bent by the application of the force. The arrangement is generally referred to as a load sell. Figure 13–4. shows an example of such a cell. This is a cylindrical tube which strain gauges have been attached. When forces are applied to the cylinder to compress it, then the strain gauge give a resistance change which is a measure of the strain hence the applied forces. Since temperature also produces a resistance change, the signal conditioning circuit used has to be able to eliminate the effects due to temperature.

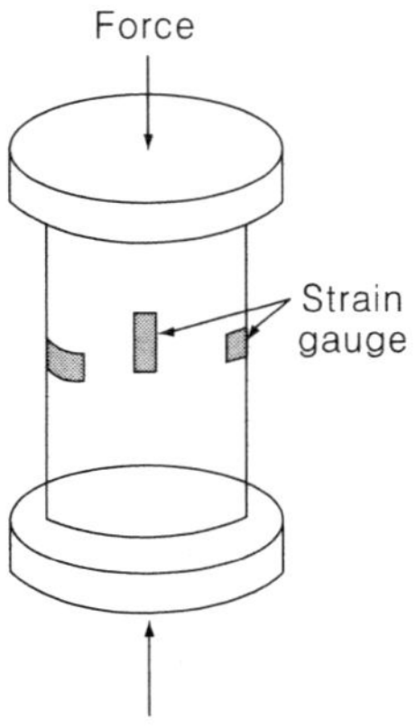

[Figure 13–4] Strain gauge and load cell

Note

- strain gauge : 스트레인게이지, 힘 혹은 압력 등의 변화를 저항변화로 변환하는 변환기의 일종
- road cell : 로드셀, 힘 혹은 무게를 측정하는 변환기
- strain : 변형
- stretch : 잡아당기다
- compress : 압축하다
- bent : 굽히다
- cylindrical tube : 원통형 튜브
- eliminate : 무시하다

5. Piezoelectric sensor

Piezoelectric materials when stretched or compressed generate electric charges with one face of the material becoming positively charged and the opposite face negatively charged. As a result, a voltage is produced. Piezoelectric materials are ionic crystals which when stretched or compressed results in the charge distribution in the crystal changing so that there is a net displacement of charge with one face of the material becoming positively charged and the other negatively charged.

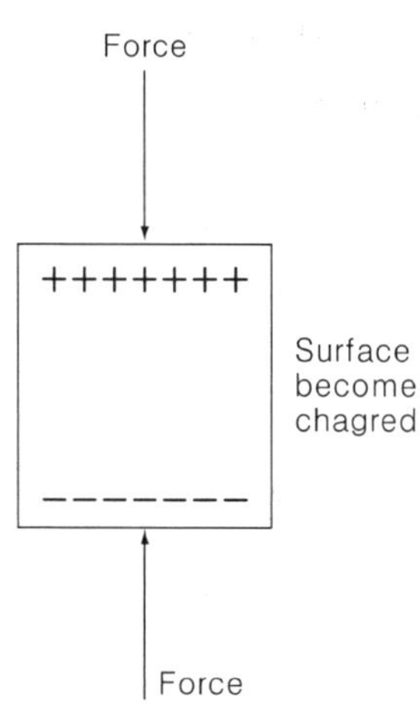

[Figure 13-5] Piezoelectricity

Piezoelectric sensors are used for the measurement of pressure, force and acceleration. The application have, however, to be such that the charge produced by the pressure does not have much time to leak off and thus tend to be used mainly for transient rather than steady pressures.

Note

- piezoelectric material : 압전재료
- electric charge : 전하
- ionic crystal : 이온 결정체
- transient pressure : 과도압력
- steady pressure : 정상압력
- leak : 누설
- tend to ~ : ~하는 경향이 있다

6. Thermocouple

If two different metals are joined together, a potential difference occurs across the junction. The potential difference depends on the metals used and the temperature of the junction. A thermocouple is a complete circuit involving two such junctions. If both junctions are at the same temperature, there is no net e.m.f. if, however there is a difference in temperature between the two junctions, there is an e.m.f. the value of this e.m.f E depend on the two metals concerned and the temperature t of both junction.

Note

- potential difference : 전위차
- junction : 접합점
- depend on ~ : ~에 의존한다
- thermocouple : 열전대, 온도를 측정하는 기전력형 센서

Actuation systems

Actuation systems are the elements of control systems which are responsible for transforming the output of a microprocessor of control system into a controlling action on a machine or device. Thus, for example, we might have an electrical output from the controller which has to be transformed into a linear or rotational motion to move a load by cylinder or motor. Another example might be where as electrical output from the controller has to be transformed into a action which controls the amount of liquid passing along a pipe

Note

- actuation system : 구동계통
- linear motion : 직선운동
- rotational motion : 회전운동
- cylinder : 실린더, 직선운동하는 장치의 명칭

Pneumatic and hydraulic system

Pneumatic signals are often used to control final control elements, even when the control system is otherwise electrical. This is because such signals can be used to actuate large valves and other high power control devices and so move significant loads. The main drawback with pneumatic system is, however, the compressibility of air. Hydraulic signals can be used for even higher power control device but are mere expensive than pneumatic systems and there are hazards associated with oil leak which do not occur with air leak.

Note

- pneumatic system : 공압 시스템
- final control element : 최종제어요소, 액추에이터의 운동방향, 속도, 힘을 제어하기 위한 요소
- drawback : 약점
- compressibility : 압축성
- hazard : 위험요소
- oil leak : 기름의 누설

Electrical system

In any discussion of electrical systems used as actuators for control, the discussion has to include:

1. Switching device such as mechanical switches, e.g. relay, or solid-state switches, thyristors, transistors, where the control signal switches on of off some electrical device, perhaps a heater or a motor
2. Solenoid type device where a current through a solenoid is used to actuate a soft iron core, as, for example, the solenoid operated hydraulic/pneumatic valve where a control current through a solenoid is used to actuate a hydraulic/pneumatic flow.
3. Drive system, such as D.C. and A.C. motors, where a current through a motor is used to produce rotation.

Note

- include : 포함한다
- relay : 릴레이, 전자석으로 작동되는 여러 개의 절연된 접점을 가진 스위치
- solid-state switch : 반도체형 스위치
- thyristor : 사이리스터, 전류 제어 기능을 가진 반도체 소자
- solenoid : 솔레노이드, 전류의 인가로 자장을 형성하여 작동하는 장치
- soft iron : 연철
- solenoid operated hydraulic/pneumatic valve : 솔레노이드 작동의 유압/공압 밸브

Mechanical system

Mechanisms are devices which can be considered to be motion converters in that they transform motion from one form to some other required form. They might, for example, transform linear motion into rotational motion, or motion in one direction into a motion in a direction at right angles, or perhaps a linear reciprocating motion into rotary motion, as in the internal combustion engine where the reciprocating motion of the piston is converted into rotation of the crank and hence the drive shaft.

Mechanical element can be included the use of linkages, cams, gears, rack-and-pinion, chains, belt drivers, etc. For example, the rack-and-pinion can be used to convert rotational motion into linear motion. Parallel shaft gears might be used to reduce a shaft speed. Bevel gears might be used for the transmission of rotary motion through 90°. A toothed belt or chain drive might be used to transform rotary motion about one axis to motion about another. Cams and linkages can be used to obtain motions which are prescribed to vary in a particular manner.

Note

- mechanism : 기계구조
- linear motion : 직선운동
- rotational motion : 회전운동
- linear reciprocating motion : 직선왕복운동
- internal combustion engine : 내연기관
- parallel shaft gear : 평행축기어
- linkage : 링크, 연쇄

Control devices

1. Relays;

The electrical relays offers a simple on/off switching action in response to a control signal, Figure 13-6. illustrates the principle. When a current flows through the coil of wire a magnetic field is produced. This pulls a movable arms, the armature, that force the contacts to open or close; usually there are two sets of contacts with one being opened and the other closed by the action(change-over contact). This action might then be used to supply a current to a motor or perhaps an electric heater in a temperature control system.

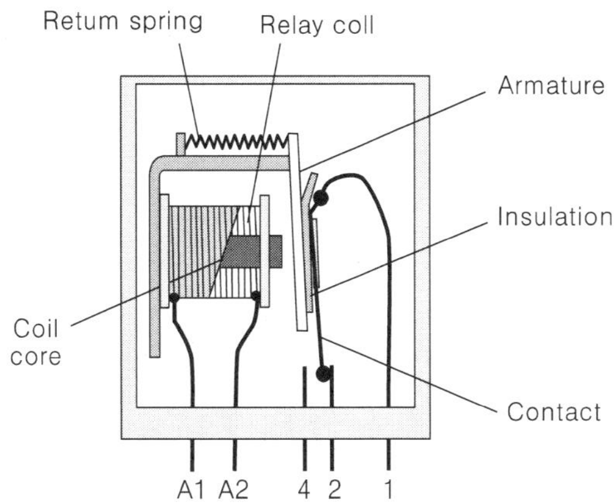

[Figure 13-6] Relay

Note

- in response to ~ : ~에 대하여
- magnetic field : 자장
- armature : 아마추어, 전기자
- contact : 접점
- change-over contact : c접점, 전환접점

2. PLC(Programmable logic controller)

a. PLC in Automation

A programmable logic controller (PLC), or programmable controller is a digital computer used for automation of industrial processes, such as control of machinery on factory assembly lines. Unlike general-purpose computers, the PLC is designed for multiple inputs and output arrangements, extended temperature ranges, immunity to electrical noise, and resistance to vibration and impact. Programs to control machine operation are typically stored in battery-backed or non-volatile memory. The facilities for extensive input/output (I/O) arrangement connect the PLC to sensors and actuators. PLCs read limit switches, analog process variables (such as temperature and pressure), and the positions of complex positioning systems. Some even use machine vision. On the actuator side, PLCs operate electric motors, pneumatic or hydraulic cylinders, magnetic relays or solenoids, or analog outputs. The input/output arrangements may be built into a simple PLC, or the PLC may have external I/O modules attached to a computer network that plugs into the PLC.

[Figure 13-7] PLC

Note

- automation : 자동화
- assembly line : 조립라인
- immunity : 면역, 내구성
- non-volatile memory : 비휘발성 메모리
- limit switche : 리밋 스위치, 접촉식으로 작동하는 센서의 일종
- complex positioning system : 복잡한 위치제어 시스템
- machine vision : 기계감시용 영상장치

b. Basic design of a PLC

The term "programmable logic controller" is defined as follows by IEC 61131, part 1:

"A digitally operating electronic system, designed for use in an industrial environment, which use a programmable memory for the internal storage of user-oriented instructions for implementing specific functions such as logic, sequencing, timing, counting and arithmetic to control through digital or analog inputs and outputs, various types of machines or processes. Both the PC and its associated peripherals are designed so that they can be easily integrated into an industrial control system and easily used in all their intended functions."

A programmable logic controller is therefore nothing more than a computer, tailored specifically for certain control tasks.

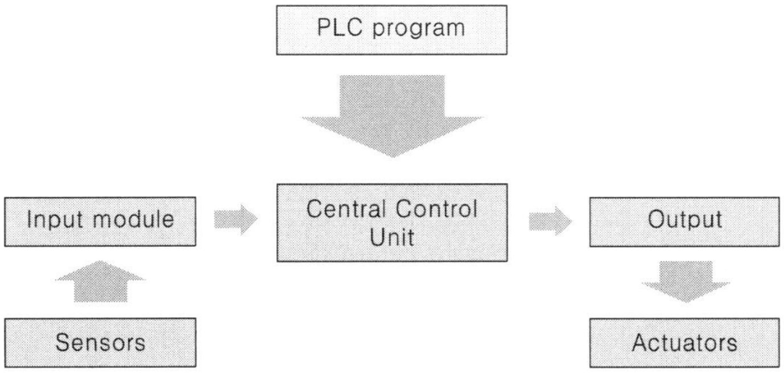

[Figure 13-8] Illustrates the system components of a PLC

The function of an input module is to convert incoming signals into signals which can be processed by the PLC and to pass these to the central control unit. The reverse task is performed by an output module. This converts the PLC signal into signals suitable for the actuators.

Note

- user oriented instruction : 사용자 정의 명령어
- sequencing : 순차제어
- timing : 시간지연
- counting : 계수
- peripheral : 주변장치
- intended function : 의도된 기능
- central control unit : CCU, 중앙처리장치

c. Programming language

IEC 61131-3 defines five programming languages. Those are 2 textual languages and 3 graphical languages. 2 textual languages are Instruction List(IL) and Structured Text(ST) and 3 graphical languages are Ladder Diagram(LD), Function Block Diagram(FBD) and Sequential Function Chart(SFC). Although the functionality and structure of these language is very different, these are treated as one language family by IEC 61131-3 with overlapping structure elements(variable declaration, organization parts such as function and function block, etc.) and configuration elements.

The languages can be mixed in any way within a PLC project. Provision has been made for future expansion, (such as function block principle or the language Structured Text) plus necessary information technology details(data type etc.) have been incorporated.

Note

- IEC 61131 : PLC 국제표준
- textual language : 문장형 언어
- graphical language : 그래픽 언어
- variable declaration : 변수 선언
- function : 출력이 기억되지 않는 특정 기능을 가진 소프트웨어 모듈
- function block : 출력이 기억되는 특정 기능을 가진 소프트웨어 모듈

14 Hydraulic Machinery

Hydraulic machinery

Hydraulic machinery is machines and tools which use fluid power to do work. Heavy equipment is a common example.

In this type of machine, high pressure hydraulic fluid is transmitted throughout the machine to various hydraulic motors and hydraulic cylinders. The fluid is controlled directly or automatically by control valves and distributed through hoses and tubes.

The popularity of hydraulic machinery is due to the very large amount of power that can be transferred through small tubes and flexible hoses, and the high power density and wide array of actuators that can make use of this power.

[Figure 14-1] Application of Hydraulic

Note

- hydraulic machinery : 유압유체기계(류)
- tool : 도구, 공구
- fluid power : 유체동력
- hydraulic motor : 유압모터
- hydraulic cylinder : 유압 실린더
- popularity : 대중성
- array : 배열
- actuator : 액추에이터

Hydraulic power

A force acting on a small area can create a much larger force by acting on a larger area by virtue of hydrostatic pressure.

Hydraulic machinery offers a very large amount of power and force with relatively small components. A typical hydraulic cylinder with a 75 mm bore, for example, can supply 89 000 N. The power transmission in a hydraulic system is easily controlled with valves.

Some parts of a hydraulic system will operate at about 20 MPa (pilot controls, vehicle brakes). The main hydraulic actuators (for example, cylinders or fluid motors) will typically operate in the range of 7000 – 42000 kPa. With advances in materials and design, there is a trend toward higher pressure, with some systems operating to 100 000 kPa and some exotic systems with titanium hardware operating at over 350 000 kPa.

Note

- hydraulics : 유압
- hydrostatic pressure : 유체 정압
- bore : 구멍, 구경(口徑)
- by virtue of : ~에 의해
- pressurized fluid : 압축 유체
- exotic system : 외국의 시스템

Force multiplication

An interesting thing about hydraulic systems is the ability to apply force multiplication. In the figure, if the smaller cylinder is one centi-meter diameter, and the larger cylinder is ten centi-meters diameter. If the force exerted on smaller cylinder is 10 kgf, the force exerted by larger cylinder is 1000 kgf because larger cylinder is a hundred times larger in area as smaller cylinder. The downside to this is that you have to move smaller cylinder a hundred centi-meters to move larger cylinder one centi-meter.

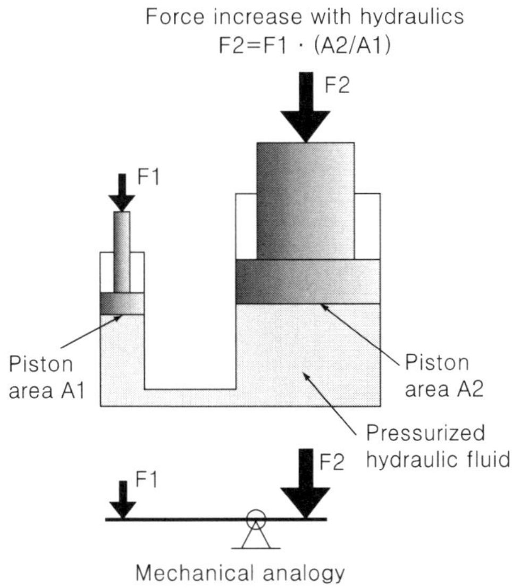

[Figure 14-2] Force multiplication

Note

- force multiplication : 힘의 증가
- diameter : 직경
- exerted : 작용되는

Directional control valve

Directional control valves route the fluid to the desired actuator. They usually consist of a spool inside a cast iron or steel housing. The spool slides to different positions in the housing, intersecting grooves and channels route the fluid based on the spool's position.

Sliding the spool to one side routes the hydraulic fluid to an actuator and provides a return path from the actuator to tank. When the spool is moved to the opposite direction the supply and return paths are switched.

The spool position may be actuated by mechanical levers, hydraulic pilot pressure, or solenoids which push the spool left or right.

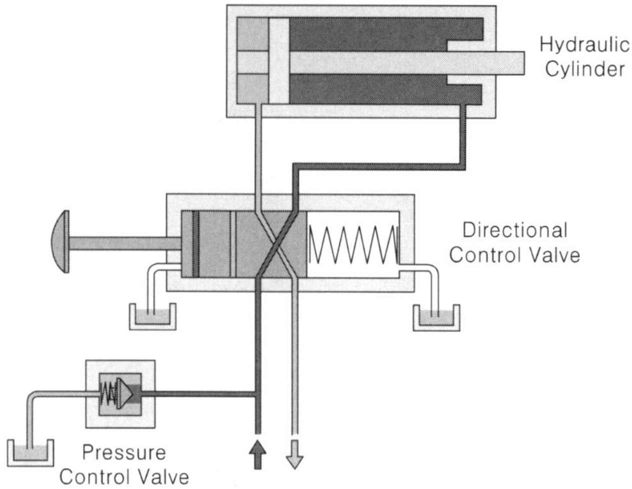

[Figure 14-3] Construction of Hydraulic system

Note

- directional control valve : 방향제어밸브
- spool : 스풀, 밸브내에서 유로를 결정하는 장치
- cast iron : 주철
- route : 보내다
- housing : 덮개, 외형
- hydraulic fluid : 유압작동유
- hydraulic pilot pressure : 간접작동의 유압 압력
- hydraulic circuit : 유압회로

Pressure control valve

Pressure control valve have the task of controlling and regulating the pressure in the hydraulic system and in parts of the system. There are two different kinds of valves. One is pressure relief valve and the other one is pressure reducing valve.

Pressure relief valves are used in several places in hydraulic machinery; on the return circuit to maintain a small amount of pressure for brakes, pilot lines, etc.. On hydraulic cylinders, to prevent overloading and hydraulic line/seal rupture. In this case, a pressure relief valve is termed a safety valve when it is attached to the pump and the valve setting is fixed at the maximum pump pressure. It only opens in case of emergency. Pressure relief valve is also used sequence valve. Sequence valves control the sequence of hydraulic circuits; to insure that one hydraulic cylinder is fully extended before another starts its stroke.

Pressure reducing valves reduce the supply pressure as needed for various circuits. They are only used to good effect in systems where a number of different pressures are required.

Note

- pressure relief valve : 압력릴리프밸브
- pressure reducing valve : 감압밸브
- return circuit : 복귀(귀환) 회로
- prevent : 방지하다
- overloading : 과부하
- rupture : 파열
- reservoir : 저장탱크
- sequence valve : 시퀀스 밸브, 순차작동의 기능을 가진 밸브

Flow control valve

Flow control valves are used to reduce the speed of cylinder or the r.p.m. of a motor. Since both values are dependent on the flow rates. For reducing the speed, the flow rates must be reduced. However, fixed displacement pumps supply a uniform flow rate. Reduction in the rate of flow supplied to the drive element is achieved according to the following principle;

A reduction in the flow cross-section in the flow control valve causes an increase in pressure ahead of this. This pressure causes the pressure relief valve to open and, consequently, result in a division of the flow rate. This division of the flow rate causes the flow volume required for the r.p.m. or speed to flow to the power component and the excess delivery to be discharged via pressure relief valve. This results in a considerable energy loss.

Note

- flow control valve : 유량제어밸브
- r.p.m : revolution per minute, 분당 회전수
- flow rate : 유량, 단위시간당 체적 • cross-section : 단면
- fixed displacement pump : 고정용량형 펌프, 펌프의 토출량이 항상 일정한 펌프
- consequently : 결과적으로
- pressure relief valve : 압력 릴리프 밸브

Hydraulic pump

Hydraulic pumps supply fluid to the components in the system. Pressure in the system develops in reaction to the load. Hence, a pump rated for 300 bar is capable of maintaining flow against a load of 300 bar.

1. **Gear pump**: This is fixed displacement pumps since the displaced volume which is determined by the tooth gap is not adjustable. The gear pump shown in the diagram is in section. The suction area S is connected to the reservoir. The gear pump operates according to the following principle;
   On gear is connected to the drive, the other is turned by the meshing teeth. The increase in volume which is produced when a tooth moves out of a mesh causes a vacuum to be generated in the suction area. The hydraulic fluid fills the tooth gaps and is conveyed externally around the housing into pressure area P. The hydraulic fluid is then forced out of the tooth gaps by the meshing of teeth and displaced into the line.

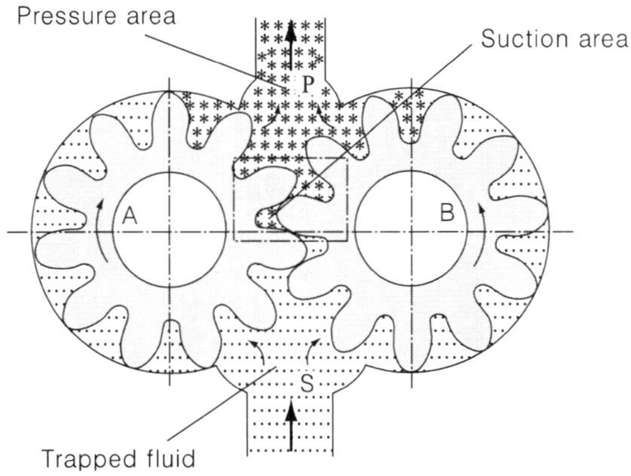

[Figure 14-4] Gear pump

Note

- hydraulic pump: 유압펌프
- be capable of ~ : ~할 능력이 있는
- fixed displacement pump : 고정용량형 펌프
- displaced volume : 배제용적
- meshing teeth : 기어의 맞물림
- suction area : 흡입측

2. **Vane pump**: Figure illustrates the operation of a vane pump. The rotor, which contains radial slots, is splined to the drive shaft and rotates inside a cam ring. Each slot contains a vane designed to mate with the surface of the cam ring as the rotor turns. Centrifugal force keeps the vanes out against the surface of the cam ring. During the first half revolution of rotor rotation, the volume increases between the rotor and cam ring. The resulting volume expansion causes a reduction of pressure. This is the suction process, which causes fluid to flow through the inlet port and fill the void. As the rotor rotates through the second half revolution, the surface of the cam ring pushes the vanes back into their slots, and the trapped volume is reduced. This positively ejects the trapped fluid through the discharge port.

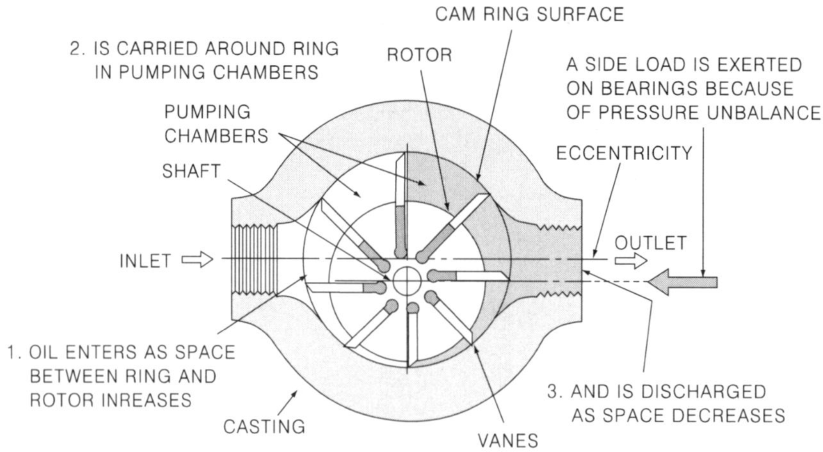

[Figure 14-5] Vane pump

Note

- vane pump : 베인펌프
- mate : 일치시키다
- suction process : 흡입과정
- discharge port : 토출구
- radial slot : 반경방향의 홈
- centrifugal force : 원심력
- trapped fluid : 토출 유량

3. **Piston pump**: A piston pump works the principle that a reciprocating piston can draw in fluid when it retracts in a cylinder bore and discharge it when it extends. The basic question is how to mechanize a series of reciprocating pistons. There are two basic types of piston pumps. One is the axial design, having pistons that are parallel to the axis of the cylinder block. Axial piston pumps can be either of the bent axis configuration or of the swash plates design. The second type of piston pump is the radial design, which has pistons arranged vertically in a cylinder block.

Note

- piston pump : 피스톤 펌프
- draw : 빨아들이다
- retract : 원위치하다
- extend : 전진운동하다
- parallel : 평행한
- axial piston pump : 축류형 피스톤 펌프
- bent axis configuration : 사축식
- swash plate : 회전사판
- reciprocate : 왕복운동을 하다
- vertically : 수직으로

Reservoir

The hydraulic fluid reservoir holds excess hydraulic fluid to accommodate volume changes from: cylinder extension and contraction, temperature driven expansion and contraction, and leaks. The reservoir is also designed to aid in separation of air from the fluid and also work as a heat accumulator to cover losses in the system when peak power is used. Design engineers are always pressured to reduce the size of hydraulic reservoirs, while equipment operators always appreciate larger reservoirs.

Some designs include dynamic flow channels on the fluid's return path that allow for a smaller reservoir.

Note

- hydraulic fluid : 유압작동유
- reservoir : 탱크
- excess : 초과의
- accommodate : 조절하다
- extension and contraction : 전진운동과 후진운동
- expansion and contraction : 팽창과 수축
- accumulator : 축열기
- channe : 경로

Operation of hydraulic cylinder

Hydraulic cylinders get their power from pressurized hydraulic fluid, which is typically oil. The cylinder consists of a cylinder barrel, in which a piston connected to a piston rod is moving. The barrel is closed by the cylinder bottom and by the cylinder head where the piston rod comes out of the cylinder. The piston has sliding rings and seals. The piston divides the inside of the cylinder in two chambers, the bottom chamber and the piston rod side chamber. The hydraulic pressure acts on the piston to do linear work.

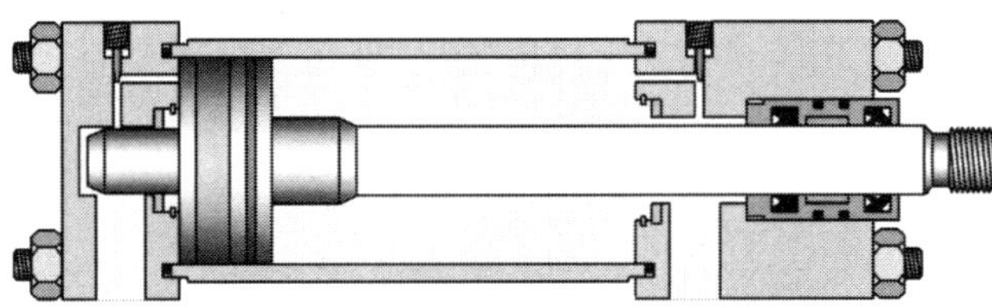

[Figure 14-6] Hydraulic cylinder

A hydraulic cylinder is the actuator or "motor" side of this system. The "generator" side of the hydraulic system is the hydraulic pump, that brings a fixed or regulated flow of oil into the system. Mounting brackets or clevises are mounted to the cylinder bottom as well as the piston rod.

Note

- hydraulic fluid : 유압작동유
- consist of ~ : ~로 구성되다
- cylinder barrel : 실린더 배럴
- mounting bracket : 마운팅 장치, 실린더의 고정 장치
- clevis : 클레비스, 실린더의 요동운동을 허용하는 마운팅 장치
- regulated : 조정된

Basic efficiency of hydraulic machine

The efficiency of hydraulic machine can be expressed in the form;

$$efficiency = \frac{energy\,output}{energy\,input} = \frac{energy\,input - energy\,loss}{energy\,input}$$

In hydraulic machines, the input and output are usually stated in terms of power or horse-power, and on this basic it is the gross or overall efficiency that concern us. It is denoted by the symbol η. The mechanical energy input or output will be represented by the horse-power delivered to or taken from the revolving shaft or other element of the machine.

Note

- efficiency : 효율
- horse-power : 마력
- overall efficiency : 전효율

15 Automatic Control

Open-loop and closed-loop control system

1. Introduction

Consider a motor. With an open-loop control system the speed of rotation of the shaft might be determine solely by the initial setting of a knob which affect the voltage applied to the motor. Any changes in the supply voltage, the characteristics of the motor as a result of temperature changes, or the shaft load will change the shaft speed but not be compensated for. There is no feedback loop. With a closed-loop control system, however, the initial setting of the control knob will be for particular shaft speed and this will be maintained by feedback, regardless of any changes in supply voltage, motor characteristics or load. In an open-loop control system, the output from the system has no effect on the input signal. In a closed-loop control system, the output dose have an effect on the input signal, modifying it to maintain an output signal at the required value.

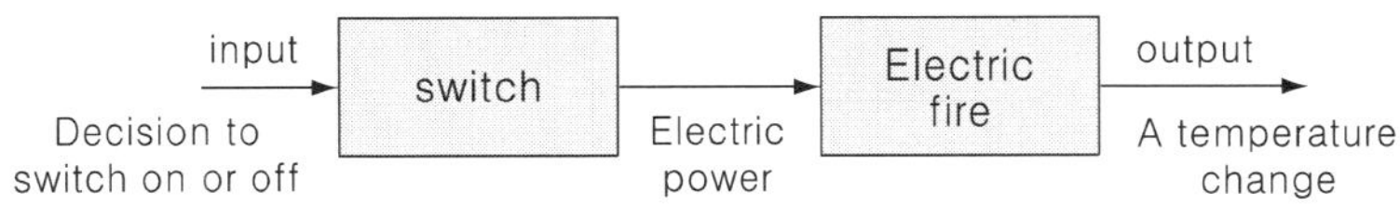

(a) Open-loop control system

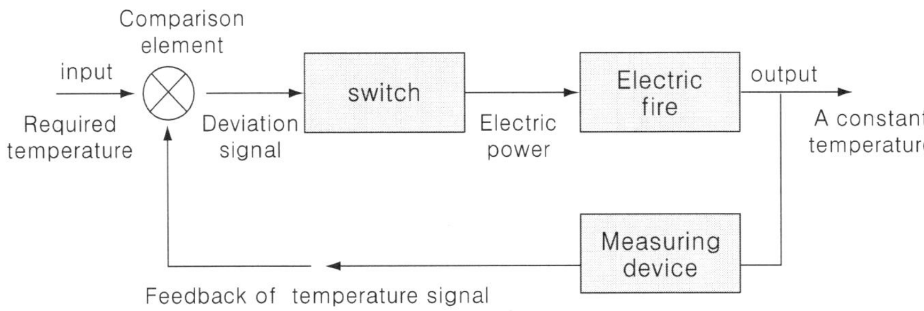

(b) Close-loop control system

[Figure 15-1] Control of heating a room

Note

- open-loop : 개(開)회로
- solely : 단지
- control system : 제어계
- feedback loop : 귀환회로
- input signal : 입력신호
- closed-loop : 폐(閉)회로
- initial setting : 초기 설정값
- knob : (계기판 등의) 손잡이
- regardless of ~ : ~에 관계없이
- required value : 요구값

2. Characteristics

Open-loop control systems have the advantage of being relatively simple and consequently low cost with generally good reliability. However, they are often inaccurate since there is no correction for error. Closed-loop control systems have the advantage of being relatively accurate in matching the actual to the required values. They are, however, more complex and so more costly with a greater chance of breakdown as a consequence of the greater number of components.

The purpose of the automatic controller for closed-loop control system is to reduce the actuating signal to a minimum, and preferably to eliminate it, with the least delay possible. The means of effecting this is the controller output signal which is transmitted to a final control element. The automatic control system includes the process. In a more specific sense, however, the control system may by considered as the combination of measuring means, controller and final control element, disregarding process.

Note

- advantage : 장점, 이점
- inaccurate : 부정확한
- correction for error : 오류에 대한 교정
- consequence : 결과
- breakdown : 고장
- automatic controller : 자동제어장치
- eliminate : 제거하다
- least delay : 최소의 지연
- measuring device : 계측기기
- actuating signal : 동작신호
- final control element : 최종 제어요소

Linear control

Linear control systems use linear negative feedback to produce a control signal mathematically based on other variables, with a view to maintaining the controlled process within an acceptable operating range.

The output from a linear control system into the controlled process may be in the form of a directly variable signal, such as a valve that may be 0 or 100% open or anywhere in between. Sometimes this is not feasible and so, after calculating the current required corrective signal, a linear control system may repeatedly switch an actuator, such as a pump, motor or heater, fully on and then fully off again, regulating the duty cycle using pulse-width modulation.

Note

- linear control : 선형제어
- variable : 변수
- operating range : 동작 범위
- with a view to ~ : ~을 예상하여

- feasible : 가능한
- repeatedly : 반복적으로
- fully off : 전폐, 완전히 닫힘
- pulse-width modulatin : 펄스 폭 제어
- corrective signal : 정정신호
- fully on : 전개, 완전히 열림
- duty cycle : 듀티 사이클

Set point

The set point is the position to which the control-point setting mechanism is fixed. The automatic controller possesses a set point scale, the set point is the scale reading translated into units of the controlled variable. Where a setting scale is not provided, the set point is the position of the control-point setting mechanism translated into units of the controlled variable. The set point can be varied manually or automatically.

Note

- set point : 설정치
- possess : 가지고 있다
- controlled variable : 제어변수
- automatic controller : 자동제어장치
- scale : 기준
- manually : 수동으로

deviation

The permanent deviation of the controlled variable from the set point is the offset. It is convenient to speak of the value for which the controller maintains as control point. The control point is the set point plus of minus offset.

Note

- deviation : 제어편차
- offset : 오프셋(차감)
- convenient : 편리한

Step function response of a process

To determine the characteristic of dynamics behavior of a process or control component, an input is artificially applied, and the output response that follows is observed. For the purpose of comparison, the input signal that is applied must follow a defined pattern. Three typical patterns of input signals are the step function, the ramp function and the sinusoidal function.

Note

- step function response : 계단함수 응답
- dynamic behavior : 동적거동
- ramp function : 램프함수
- sinusoidal function : 정현(正弦)함수

Impulse signal

An impulse is defined as a signal that has an infinite magnitude and an infinitesimally narrow width with an area under it of one, centered at zero. An impulse can be represented as an infinite sum of sinusoids that includes all possible frequencies. This definition is really hard to use in real life, so most engineers conceptualize it to a signal that is one at zero and zero everywhere else. If an impulse is used as an input to a system, the output is known as the impulse response. The impulse response defines the system because all possible frequencies are represented in the input.

Note

- impulse signal : 임펄스 신호
- infinite magnitude : 무한한 크기
- infinitesimally : 극소로
- sinusoid : 사인(sine) 곡선
- all possible frequenciy : 모든 가능한 주파수
- conceptualize : 개념화시키다

Domain

1. **Time domain** : This is the domain that most people are familiar with. A plot in the time domain shows the magnitude of a signal at a point in time.
2. **Frequency domain** : This is the domain that engineers are glad exists. It's unfamiliar to most people, but makes the math associated with analog signal processing much easier than if it's analyzed in the time domain. A plot in the frequency domain shows either the phase shift or magnitude of a signal at each frequency that it exists at. These can be found by taking the Fourier transform of a time signal and are plotted similarly to a bode plot.

Note

- domain : (수학 등의)변역, 정의역
- unfamiliar : 익숙하지 않은
- analog signal processing : 아날로그 신호처리
- phase shift : 위상 변화(이동)
- magnitude : 크기
- Fourier transform : 퓨리에 변환
- bode plot : 보드선도

Time constant

The time constant is defined in many different ways. Its magnitude is always the same. Thus, one definition of the time constant is that it expresses the time which the controlled variable needs to pass through 63.2% of its total change. No matter how big is the step input for a given process, the controlled variable always reach 63.2% of its final value within the same amount of time which is given by the time constant T.

Note

- time constant : 시간 상수
- magnitude : 크기
- controlled variable : 제어량, 제어하는 일이 목적으로 되어 있는 양
- no matter how ~ : 아무리 ~하더라도

Laplace transform

The Laplace transform has many important applications in mathematics, physics, optics, electrical engineering, control engineering, signal processing, and probability theory. In mathematics, it is used for solving differential and integral equations. In physics, it is used for analysis of linear time-invariant systems such as electrical circuits, harmonic oscillators, optical devices, and mechanical systems. In this analysis, the Laplace transform is often interpreted as a transformation from the time-domain, in which inputs and outputs are functions of time, to the frequency-domain, where the same inputs and outputs are functions of complex angular frequency, or radians per unit time. Given a simple mathematical or functional description of an input or output to a system, the Laplace transform provides an alternative functional description that often simplifies the process of analyzing the behavior of the system, or in synthesizing a new system based on a set of specifications.

Note

- Laplace transform : 라플라스 변환
- optics : 광학
- signal process : 신호처리
- probability theory : 확률 이론
- differential equation : 미분방정식
- integral equation : 적분 방정식
- linear time-invariant system : 선형 시간불변계
- harmonic oscillator : 조화 가진기
- optical device : 광학기기
- complex angular frequency : 복소 각 진동수(주파수)

PID controller

1. Introduction

A proportional-integral-derivative controller (PID controller) is a generic control loop feedback mechanism widely used in industrial control systems. A PID controller attempts to correct the error between a measured process variable and a desired setpoint by calculating and then outputting a corrective action that can adjust the process accordingly.

The PID controller calculation involves three separate parameters; the Proportional, the Integral and Derivative values. The Proportional value determines the reaction to the current error, the Integral determines the reaction based on the sum of recent errors and the Derivative determines the reaction to the rate at which the error has been changing. The weighted sum of these three actions is used to adjust the process via a control element such as the position of a control valve or the power supply of a heating element.

Note

- PID controller : 비례/미분/적분 제어기
- generic : 일반적인
- control loop feedback mechanism : 피드백 구조를 가진 제어 루프
- process variable : 처리변수
- setpoint : 설정값
- proportional value : 비례값
- integral value : 적분값
- derivate value : 미분값
- weighted : 가중치
- separate parameter : 분리 매개변수

2. Application

By "tuning" the three constants in the PID controller algorithm the PID can provide control action designed for specific process requirements. The response of the controller can be described in terms of the responsiveness of the controller to an error, the degree to which the controller overshoots the setpoint and the degree of system oscillation. Note that the use of the PID algorithm for control does not guarantee optimal control of the system or system stability.

Some applications may require using only one or two modes to provide the appropriate system control. This is achieved by setting the gain of undesired control outputs to zero. A PID controller will be called a PI, PD, P or I controller in the absence of the respective control actions. PI controllers are particularly common, since derivative action is very sensitive to measurement noise, and the absence of an integral value may prevent the system from reaching its target value due to the control action.

Note

- tuning : 조율
- responsiveness : 응답
- overshoot : (목표량을)초과하다
- oscillation : 발진
- optimal control : 최적제어
- stability : 안정성
- gain : 이득
- in the absence of ~ : ~이 없을 경우에
- sensitive : 민감한
- target value : 목표치

Nyquist plot

A Nyquist plot is used in automatic control and signal processing for assessing the stability of a system with feedback. It is represented by a graph in polar coordinates in which the gain and phase of a frequency response are plotted. The plot of these phasor quantities shows the phase as the angle and the magnitude as the distance from the origin. This plot combines the two types of Bode plot – magnitude and phase – on a single graph, with frequency as a parameter along the curve.

Assessment of the stability of a closed-loop negative feedback system is done by applying the Nyquist stability criterion to the Nyquist plot of the open-loop system (i.e. the same system without its feedback loop). This method is easily applicable even for systems with delays which may appear difficult to analyze by means of other methods.

Note

- Nyquist plot : 나이키스트 선도
- automatic control : 자동제어
- signal processing : 신호처리
- assess : 평가하다
- polar coordinate : 극좌표
- gain : 이득
- phase : 위상
- frequency response : 주파수 응답
- bode plot : 보드 선도
- assessment : 평가
- Nyquist stability criterion : 나이키스트 안정성 기준

16 Basic Electric and Electronics

Charge

"Charge" is the technical term used to indicate that an object has been prepared so as to participate in electrical forces. This is to be distinguished from the common usage, in which the term is used indiscriminately for anything electrical. For example, although we speak colloquially of "charging" a battery, you may easily verify that a battery has no charge in the technical sense, e.g. it does not exert any electrical force on a piece of tape that has been prepared as described in the previous section.

Note

- charge : 전하
- so as to : ~ 하도록
- distinguish : 구별하다
- indiscriminate : 무계획적인
- colloquially : 구어체적으로
- exert : (힘을) 가하다

Coulomb's law

A large body of experimental observations can be summarized as follows:

Coulomb's law is that the magnitude of the force acting between pointlike charged objects at a center-to-center distance r is given by the equation.

$$|F| = k\frac{|q_1 \| q_2|}{r^2}$$

Where, the constant k equals 9.0×10^9 N · m^2/C^2. The force is attractive if the charges are of different signs, and repulsive if they have the same sign.

Clever modern techniques have allowed the $1/r^2$ form of Coulomb's law to be tested to incredible accuracy, showing that the exponent is in the range from 1.9999999999999998 to 2.0000000000000002.

Note that Coulomb's law is closely analogous to Newton's law of gravity, where, the magnitude of the force is Gm_1m_2/r^2, except that there is only one type of mass, not two, and gravitational forces are never repulsive. Because of this close analogy between the two types of forces, we can recycle a great deal of our knowledge of gravitational forces.

Note

- experimental : 실험적
- observation : 관찰
- center-to-center distance : 중심거리
- attractive : 인력이 있는
- repulsive : 반발하는
- incredible : 놀라운
- exponent : x^n에 대한 누승 지수
- analogous : 유사한

Electrical current

If the fundamental phenomenon is the motion of charged particles, then how can we define a useful numerical measurement of it? We might describe the flow of a river simply by the velocity of the water, but velocity will not be appropriate for electrical purposes because we need to take into account how much charge the moving particles have, and in any case there are no practical devices sold at Radio Shack that can tell us the velocity of charged particles. Experiments show that the intensity of various electrical effects is related to a different quantity: the number of coulombs of charge that pass by a certain point per second. By analogy with the flow of water, this quantity is called the electric current, I. Its units of coulombs/second are more conveniently abbreviated as amperes, 1 A=1 C/s.

The definition is summarized as follows.

When charged particles are exchanged between regions of space A and B, the electric current flowing from A to B is defined as $I = \triangle q / \triangle t$, where $\triangle q$ is the change in the total charge of region B.

Note

- current : 전류
- numerical : 숫자상의
- appropriate : 적합한
- Radio Shack : 무선통신실
- analogy : 유사, 상사
- abbreviate : 단축하다
- phenomenon : 현상
- simply : 간단하게
- take into account ~ : ~을 고려하다.
- intensity : 강도
- conveniently : 편리하게

Voltage

Voltage (sometimes also called electric or electrical tension) is the difference of electrical potential between two points of an electrical or electronic circuit, expressed in volts. It measures the potential energy of an electric field to cause an electric current in an electrical conductor. Depending on the difference of electrical potential it is called extra low voltage, low voltage, high voltage or extra high voltage.

Note

- voltage : 전압
- tension : 응력
- difference of electrical potential : 전위차
- potential energy : 전위의 에너지
- electric field : 전계
- electric current : 전류
- electrical conductor : 전도체

Electrical resistance

An electron traveling through the wires and loads of the external circuit encounters resistance. Resistance is the hindrance to the flow of charge. For an electron, the journey from terminal to terminal is not a direct route. Rather, it is a zigzag path which results from countless collisions with fixed atoms within the conducting material. The electrons encounter resistance - a hindrance to their movement. While the electric potential difference established between the two terminals encourages the movement of charge, it is resistance which discourages it. The rate at which charge flows from terminal to terminal is the result of the combined affect of these two quantities

Note

- resistance : 저항
- electron : 전자
- encounter : 대항하다
- hindrance : 장애물
- charge : 전하
- terminal : 단자
- rather : 오히려
- collision : 충돌
- countless : 셀 수 없을 정도로 많은
- atom : 원자
- conducting material : 전도체
- electric potential difference : 전위차
- encourage : 촉진하다

Ohm's law

Electrical resistance shares some conceptual parallels with the mechanical notion of friction. The SI unit of electrical resistance is the ohm, symbol Ω. The resistance of an object determines the amount of current through the object for a given potential difference across the object.

$$R = \frac{V}{I}$$

where

R is the resistance of the object, measured in ohms, equivalent to $J \cdot s/C^2$.

V is the potential difference across the object, measured in volts.

I is the current through the object, measured in amperes.

For a wide variety of materials and conditions, the electrical resistance does not depend on the amount of current through or the amount of voltage across the object, meaning that the resistance R is constant.

Note

- electrical resistance : 전기 저항
- share : 공유하다
- friction : 마찰
- amount : 양
- potential difference : 전위차
- equivalent : 동등한

Direct current

Direct current(DC or "continuous current") is the unidirectional flow of electric charge. Direct current is produced by such sources as batteries, thermocouples, solar cells, and commutator-type electric machines of the dynamo type. Direct current may flow in a conductor such as a wire, but can also be through semiconductors, insulators, or even through a vacuum as in electron or ion beams. In direct current, the electric charges flow in the same direction, distinguishing it from alternating current(AC). A term formerly used for direct current was Galvanic current.

Direct current may be obtained from an alternating current supply by use of a current-switching arrangement called a rectifier, which contains electronic elements(usually) or electromechanical elements (historically) that allow current to flow only in one direction. Direct current may be made into alternating current with an inverter or a motor-generator set.

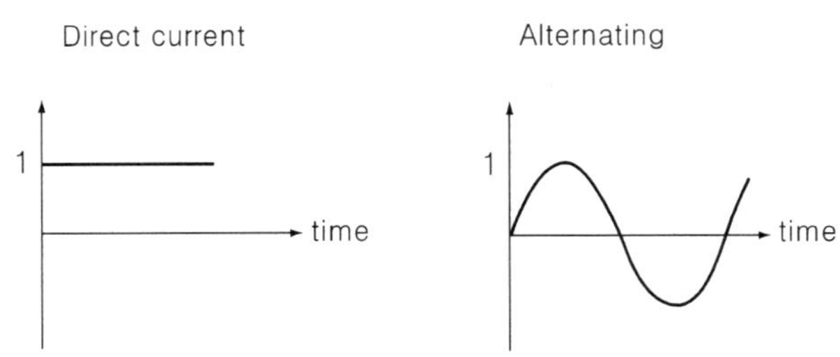

[Figure 16-1] Direct current and Alternating current

Note

- direct current : 직류
- electric charge : 전하
- solar cell : 태양전지
- dynamo : 발전기
- semiconductor : 반도체
- alternating current : 교류
- inverter : 인버터, 주파수 변환기
- unidirectional : 단일 방향성의
- thermocouple : 열전대
- commutator : 정류자
- conductor : 도체
- insulator : 절연체
- rectifier : 정류기

Alternating current

An alternating current (AC) is an electrical current whose magnitude and direction vary cyclically, as opposed to direct current, whose direction remains constant. The usual waveform of an AC power circuit is a sine wave, as this results in the most efficient transmission of energy. However in certain applications different waveforms are used, such as triangular or square waves.

Used generically, AC refers to the form in which electricity is delivered to businesses and residences. However, audio and radio signals carried on electrical wire are also examples of alternating current. In these applications, an important goal is often the recovery of information encoded (or modulated) onto the AC signal.

Note

- alternating current : 교류
- magnitude : 크기
- cyclically : 주기적으로
- waveform : 파형
- efficient : 효율적인
- transmission : 전송
- businesses : 업무용
- residence : 주거용
- encode : 부호화하다.
- modulate : 변조하다.

Kirchhoff's Current Law

Kirchhoff's Current Law, also known as Kirchhoff's Junction Law and Kirchhoff's First Law, defines the way that electrical current is distributed when it crosses through a junction – a point where three or more conductors meet. Specifically, the law states that the algebraic sum of current into any junction is zero.

Since current is the flow of electrons through a conductor, it cannot build up at a junction, meaning that current is conserved: what comes in must come out. When performing calculations, current flowing into and out of the junction typically have opposite signs.

This allows Kirchhoff's Current Law to be restated as:

The sum of current into a junction equals the sum of current out of the junction.

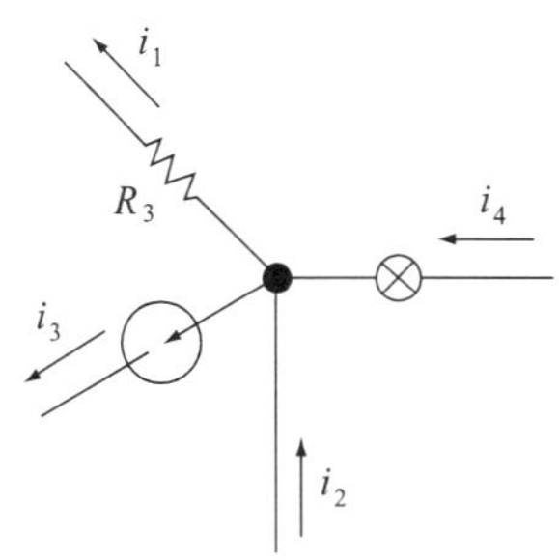

[Figure 16-2] Diagram of Kirchhoff's current law

In the figure. a junction of four conductors (i.e. wires) is shown. The currents i_2 and i_3 are flowing into the junction, while i_1 and i_4 flow out of it. In this example, Kirchhoff's Junction Rule yields the following equation:

$$i_2 + i_3 = i_1 + i_4$$

Note

- Kirchhoff's Current Law : 키르히호프의 전류법칙
- algebraic sum : 대수합
- since : 왜냐하면
- electron : 전하
- conductor : 도체
- conserve : 보존하다
- junction : 접합점
- yield : 생기게 하다

Kirchhoff's Vlotage Law

Kirchhoff's Voltage Law describes the distribution of voltage within a loop, or closed conducting path, of an electrical circuit. Specifically, Kirchhoff's Voltage Law states that the algebraic sum of the voltage (potential) differences in any loop must equal zero.

The voltage differences include those associated with electromagnetic fields (emfs) and resistive elements, such as resistors, power sources (i.e. batteries) or devices (i.e. lamps, televisions, blenders, etc.) plugged into the circuit.

Kirchhoff's Voltage Law comes about because the electrostatic field within an electric circuit is a conservative force field. As you go around a loop, when you arrive at the starting point has the same potential as it did when you began, so any increases and decreases along the loop have to cancel out for a total change of 0. If it didn't, then the potential at the start/end point would have two different values.

Note

- Kirchhoff's Voltage Law : 키르히호프의 전압법칙
- describe : 설명하다
- distribution : 분포
- state : 주장하다
- algebraic sum : 대수합
- electrostatic field : 정전기 계
- conservative : 전통적인

Voltmeter

A voltmeter is an instrument used for measuring the electrical potential difference between two points in an electric circuit. Analog voltmeters move a pointer across a scale in proportion to the voltage of the circuit; digital voltmeters give a numerical display of voltage by use of an analog to digital converter.

Small portable instruments, usually equipped with facilities to also measure current and resistance in the form of a multimeter, are standard test instruments used in electrical and electronics work.

General purpose analog voltmeters may have an accuracy of a few percent of full scale, and are used with voltages from a fraction of a volt to several thousand volts. Digital meters can be made with high accuracy, typically better than 1%. Specially calibrated test instruments have higher accuracies, with laboratory instruments capable of measuring to accuracies of a few parts per million. Meters using amplifiers can measure tiny voltages of microvolts or less.

[Figure 16－3] Digital multimeter

Note

- voltmeter : 전압계
- instrument : 계기
- electrical potential difference : 전기적 전위차
- in proportion to～ : ～에 비례하여
- analog to digital converter : A/D변환기
- multimeter : 멀티미터, 전압 · 전류 · 저항을 모두 측정할 수 있는 계기
- fraction : 아주조금
- accuracy : 정확도
- calibrate : 눈금을 조정하다
- tiny : 아주 작은

Ammeter

This is instrument for measuring either direct or alternating electric current, in amperes. An ammeter can meásure a wide range of current values because at high values only a small portion of the current is directed through the meter mechanism; a shunt in parallel with the meter carries the major portion.

Note

- instrument : 계기
- direct or alternating electric current : 직류 혹은 교류 전류
- ammeter : 전류계
- portion : 부분
- shunt : 분기 회로를 만들다.

Semiconductor

Semiconductors intrinsic electrical properties are often permanently modified by introducing impurities by a process known as doping. Usually, it is sufficient to approximate that each impurity atom adds one electron or one "hole" that may flow freely. Upon the addition of a sufficiently large proportion of impurity dopants, semiconductors will conduct electricity nearly as well as metals. Depending on the kind of impurity, a doped region of semiconductor can have more electrons or holes, and is named N-type or P-type semiconductor material, respectively. Junctions between regions of N- and P-type semiconductors create electric fields, which cause electrons and holes to be available to move away from them, and this effect is critical to semiconductor device operation. Also, a density difference in the amount of impurities produces a small electric field in the region which is used to accelerate non-equilibrium electrons or holes.

Note

- semiconductor : 반도체
- intrinsic : 진성의
- impuritiy : 불순물
- doping : 도핑, 반도체에 불순물을 첨가하여 전기적 특성을 얻는 것.
- dopant : 반도체에 첨가하는 미세한 불순물
- approximate : ~에 가까워지다.
- hole : 정공
- electron : 전자
- conduct : 전도하다
- respectively : 각각
- electric field : 전계
- density : 밀도
- non-equilibrium : 불평형

Diode

The P-N junction diode is a device made from a P-N junction. At the junction of a P-type and an N-type semiconductor there forms a region called the depletion zone which blocks current conduction from the N-type region to the N-type region, but allows current to conduct from the P-type region to the N-type region. Thus when the device is forward biased, with the P-side at higher electric potential, the diode conducts current easily; but the current is very small when the diode is *reverse biased.*

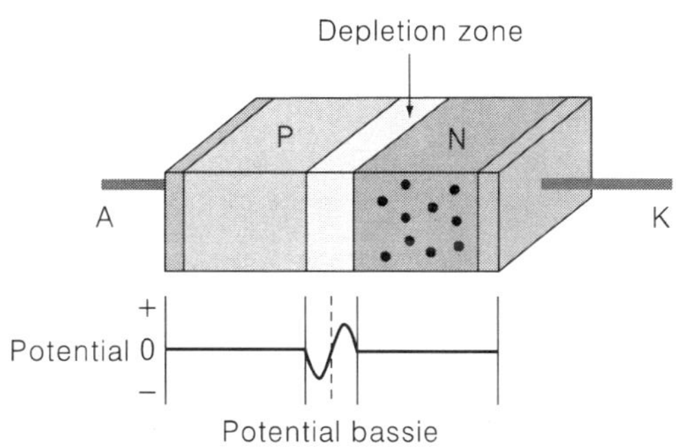

[Figure 16-4] Structure of Diode

Note

- diode : 다이오드, 전류가 한 방향으로만 흐른다
- junction : 접합
- depletion zone : 공핍층
- forward bias : 순방향 바이어스
- electric potential : 전위
- reverse bias : 역방향 바이어스
- potential barrie : 전위장벽

Transistor

Bipolar junction transistors are formed from two P-N junctions, in either N-P-N or P-N-P configuration. The middle, or base, region between the junctions is typically very narrow. The other regions, and their associated terminals, are known as the emitter and the collector. A small current injected through the junction between the base and the emitter changes the properties of the base-collector junction so that it can conduct current even though it is reverse biased. This creates a much larger current between the collector and emitter, controlled by the base-emitter current.

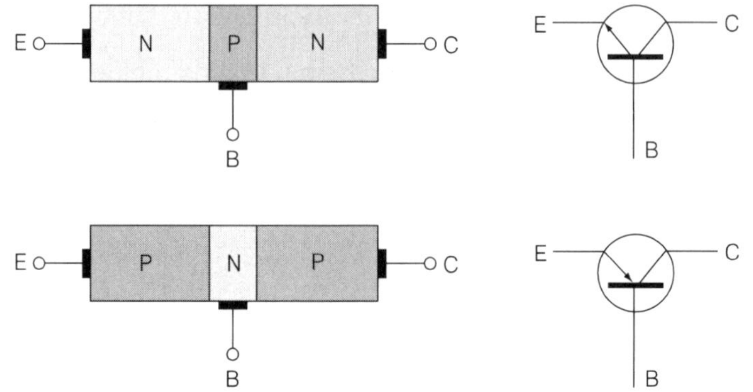

[Figure 16-5] Structure of transistor

Note

- bipolar junction transistor : BJT, 3개의 단자를 가진 트랜지스터로 스위칭 혹은 증폭기능이 있다.
- configuration : 배치
- terminal : 단자
- inject : 주입하다.
- junction : 접합
- conduct : 전도하다
- reverse bias : 역방향 바이어스

Usage of transistor

Switches

Transistors are commonly used as electronic switches, for both high power applications including switched-mode power supplies and low power applications such as logic gates.

Amplifiers

From mobile phones to televisions, vast numbers of products include amplifiers for sound reproduction, radio transmission, and signal processing. The first discrete transistor audio amplifiers barely supplied a few hundred milliwatts, but power and audio fidelity gradually increased as better transistors became available and amplifier architecture evolved.

Transistors are commonly used in modern musical instrument amplifiers, in which circuits up to a few hundred watts are common and relatively cheap. Transistors have largely replaced valves (electron tubes) in instrument amplifiers. Some musical instrument amplifier manufacturers mix transistors and vacuum tubes in the same circuit, to utilize the inherent benefits of both devices.

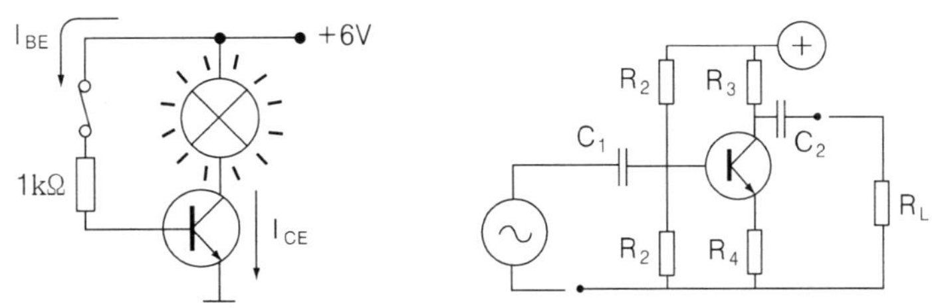

(a) Transistor for switch (b) Trnasistor for amplifier

[Figure 16-6] Usage of transistor

Note

- transistor : 트랜지스터
- logic gate : 논리게이트
- discrete : 이산적
- fidelity : 충실도
- electron tube : 전자관, 진공관의 일종
- vacuum tube : 진공관
- commonly : 일반적으로
- amplifier : 증폭기
- barely : 겨우
- evolve : 진화하다
- inherent : 고유의

17 Introduction to Micro/Nano Technology

Definition of MEMS

Micro electro-mechanical systems (MEMS) is the technology of the very small, and merges at the nano-scale into nanoelectromechanical systems (NEMS) and nanotechnology. MEMS are also referred to as micromachines in Japan, or Micro Systems Technology - MST in Europe. MEMS are separate and distinct from the hypothetical vision of Molecular nanotechnology or Molecular Electronics. MEMS are made up of components between 1 to 100 micrometers in size (i.e. 0.001 to 0.1 mm) and MEMS devices generally range in size from 20 micrometers (20 millionth of a meter) to a millimeter (thousandth of a meter). They usually consist of a central unit that processes data, the microprocessor and several components that interact with the outside such as micro-sensors.

Note

- merge : 병합하다
- hypothetical : 가상적
- micromachine : 100×10^{-9}에서 100×10^{-6}크기의 집적된 기계적 객체
- distinct from : ~와 다르게
- molecular nanotechnology : 기능을 가진 분자크기의 기계적 시스템에 대한 공학적 개념
- molecular electronics : 극소전자공학
- central unit : 중앙 장치
- interact : 상호작용하다

Microelectromechanical systems description

MEMS technology can be implemented using a number of different materials and manufacturing techniques; the choice of which will depend on the device being created and the market sector in which it has to operate.

1. Silicon

Silicon is the material used to create most integrated circuits used in consumer electronics in the modern world. The economies of scale, ready availability of cheap high-quality materials and ability to incorporate electronic functionality make silicon attractive for a wide variety of MEMS applications. Silicon also has significant advantages engendered through its material properties. In single crystal form, silicon is an almost perfect Hookean material, meaning that when it is flexed there is virtually no hysteresis and hence almost no energy dissipation. As well as making for highly repeatable motion, this also makes silicon very reliable as it suffers very little fatigue and can have service lifetimes in the range of billions to trillions of cycles without breaking. The basic techniques for producing all silicon based MEMS devices are deposition of material layers, patterning of these layers by photolithography and then etching to produce the required shapes.

Note

- implement : 수단을 주다
- silicon : 규소
- integrated circuit : 집적회로
- consumer electronics : 가전제품
- variety : 다양성
- significant : 중대한
- engender : 생기게 하다.
- single crystal form : 단결정형상
- Hookean material : 후크의 법칙을 만족하는 탄성이 있는 물체
- flex : 구부리다
- virtually : 실질적으로
- hysteresis : 이력현상
- dissipation : 분해

• fatigue : 피로
• deposition : 적층
• photolithography : 반도체 제조공정 중 하나인 리소그래피 공정
• etching : 반도체 제조공정 중 하나인 에칭 공정

2. Polymers

Even though the electronics industry provides an economy of scale for the silicon industry, crystalline silicon is still a complex and relatively expensive material to produce. Polymers on the other hand can be produced in huge volumes, with a great variety of material characteristics. MEMS devices can be made from polymers by processes such as injection moulding, embossing or stereolithography and are especially well suited to microfluidic applications such as disposable blood testing cartridges.

Note

• even though : 비록 ~할지라도
• silicon : 규소
• crystalline silicon : 투명 실리콘
• relatively : 상대적으로
• polymer : 폴리머, 중합체
• on the other hand : 이와 반대로
• a great variety of : 많은 갖가지의
• injection moulding : 사출성형
• stereolithography : 빠른 가공과 제조로 높은 정밀도와 좋은 표면 조도를 가진 재료를 만드는 기술
• microfluidic : 유체의 작용, 조정, 정밀제어를 취급하는 것.
• disposable : 일회용의

3. Metals

Metals can also be used to create MEMS elements. While metals do not have some of the advantages displayed by silicon in terms of mechanical properties, when used within their limitations, metals can exhibit very high degrees of reliability.

Metals can be deposited by electroplating, evaporation, and sputtering processes.

Commonly used metals include gold, nickel, aluminum, chromium, titanium, tungsten, platinum, and silver.

Note

- in terms of : ~에 관하여
- limitation : 한계
- deposit : 침적되다
- evaporation : 증발탈수
- property : 특성
- exhibit : 나타내다
- electroplating : 전기도금
- sputtering : 증착

Nano technology

Nanotechnology is a highly multidisciplinary field, drawing from fields such as applied physics, materials science, interface and colloid science, device physics, supramolecular chemistry (which refers to the area of chemistry that focuses on the noncovalent bonding interactions of molecules), self-replicating machines and robotics, chemical engineering, mechanical engineering, biological engineering, and electrical engineering.

Examples of nanotechnology are the manufacture of polymers based on molecular structure, and the design of computer chip layouts based on surface science. Despite the promise of nanotechnologies such as quantum dots and nanotubes, real commercial applications have mainly used the advantages of colloidal nanoparticles in bulk form, such as suntan lotion, cosmetics, protective coatings, drug delivery and stain resistant clothing.

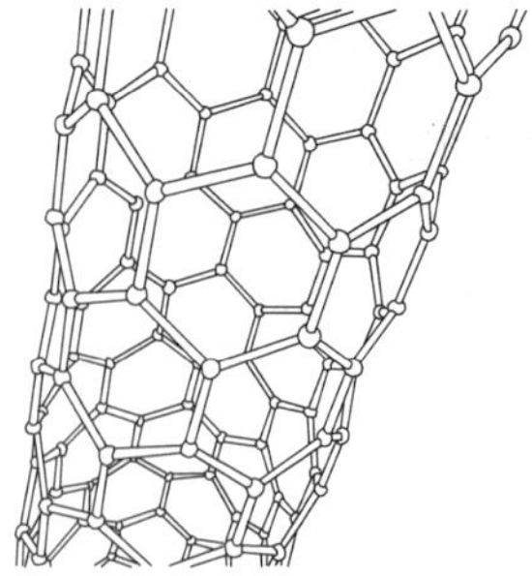

[Figure 17-1] Carbon nanotube

Note

- multidisciplinary : 융합학문
- applied physics : 응용물리학
- materials science : 재료과학
- interface and colloid science : 콜로이드와 불균질 시스템을 다루는 화학의 한 분야
- supramolecular chemistry : 고분자 화학
- noncovalent bonding : 공유결합이 아닌
- self-replicating machine : 자기복제기계
- biological engineering : 생물공학
- polymer : 폴리머, 중합체
- molecular structure : 분자구조
- despite ~ of : ~에도 불구하고
- quantum dots : 양자점
- nanotube : 나노튜브, 탄소원자가 직경 수 나노미터의 원통을 이룬 풀러렌 비슷한 원자
- colloidal : 콜로이드와 같은, 아교질의
- bulk : 용적
- stain : 얼룩

Origin of nano-technology

The first use of the concepts in 'nano-technology' was in "There's Plenty of Room at the Bottom," a talk given by physicist Richard Feynman at an American Physical Society meeting at Caltech on December 29, 1959. Feynman described a process by which the ability to manipulate individual atoms and molecules might be developed, using one set of precise tools to build and operate another proportionally smaller set, so on down to the needed scale. In the course of this, he noted, scaling issues would arise from the changing magnitude of various physical phenomena: gravity would become less important, surface tension and Van der Waals attraction would become more important, etc.

The term "nanotechnology" was defined by Tokyo Science University Professor Norio Taniguchi in a 1974 paper as follows: "'Nano-technology' mainly consists of the processing of, separation, consolidation, and deformation of materials by one atom or by one molecule."

Note

- physicist : 물리학자
- Richard Feynman : 미국의 물리학자
- American Physical Society : 미국물리학회
- Caltech : California Institute of Technology, 캘리포니아 공과대학
- manipulate: : (기계등을)다루다
- precise : 정밀한
- proportionally : 비례적으로
- scaling issue : 크기 문제
- arise : 발생하다
- gravity : 중력
- surface tension : 표면장력
- Van Der Waals attraction : 반데르발스 인력
- consolidation : 통합
- deformation : 변형

Development of nanotechnology

Nanotechnology and nanoscience got started in the early 1980s with two major developments; the birth of cluster science and the invention of the scanning tunneling microscope (STM). This development led to the discovery of fullerenes in 1986 and carbon nanotubes a few years later. In another development, the synthesis and properties of semiconductor nanocrystals was studied. This led to a fast increasing number of metal oxide nanoparticles of quantum dots. The atomic force microscope was invented six years after the STM was invented. In 2000, the United States National

Nanotechnology Initiative was founded to coordinate Federal nanotechnology research and development.

Note

- cluster science: 클러스터 과학
- invention : 발명
- STM : scanning tunneling microscope, 주사형 터널 현미경
- discovery : 발견
- fullerene : 플러렌, 탄소 원자 60개로 구성된 공 모양의 분자로 된 물질
- carbon nanotube : 탄소 나노튜브
- synthesis : 합성
- nanocrystal : 나노결정
- metal oxide : 금속산화막
- quantum dots : 양자점
- atomic force microscope : 원자간력 현미경
- coordinate : 조정하다

Nano particle

One nanometer (nm) is one billionth, or 10^{-9} of a meter. To put that scale in context, the comparative size of a nanometer to a meter is the same as that of a marble to the size of the earth. Or another way of putting it: a nanometer is the amount a man's beard grows in the time it takes him to raise the razor to his face.

A nano particle which historically has included nano powder, nanocluster, and nanocrystal is a small particle with at least one dimension less than 100 nm. This definition can be fleshed out further in order to remove ambiguity from future nano nomenclature. A nanoparticle is an amorphous or semicrystalline zero dimensional (0D) nano structure with at least one dimension between 10 and 100nm and relatively large ($\geq$ 15%) size dispersion.

Note

- one billionth : 10억분의 1
- context : 환경
- comparative size : 비교크기
- marble : 구슬
- beard : 수염
- razor : 면도날

- nanopowder : 나노분말
- ambiguity : 모호
- amorphous: 비결정의
- nano structure : 나노 구조물
- flesh : 구체화하다
- nomenclature : 명칭
- semicrystalline: 반 결정체의
- dispersion : 분포

Applications

As of April 24, 2008 The Project on Emerging Nanotechnologies claims that over 609 nanotech products exist, with new ones hitting the market at a pace of 3-4 per week. Most applications are limited to the use of "first generation" passive nanomaterials which includes titanium dioxide in sunscreen, cosmetics and some food products. Carbon allotropes used to produce gecko tape; silver in food packaging, clothing, disinfectants and household appliances; zinc oxide in sunscreens and cosmetics, surface coatings, paints and outdoor furniture varnishes; and cerium oxide as a fuel catalyst.

Note

- application : 응용
- Project on Emerging Nanotechnologies : 2005년에 설립된 나노기술의 사회적, 정치적, 공공 안전에 관한 견해를 주장하는 기관
- claim : 주장하다
- nanotech product : 나노기술 생산품
- exist : 현존하다
- pace : 속도
- passive : 수동적인
- titanium dioxide : 이산화 티탄
- sunscreen : 햇빛타기 방지제
- cosmetics : 화장품
- carbon allotrope : 탄소 동소체
- gecko tape : 게코 도마뱀의 발바닥을 응용한 테이프
- disinfectant : 살균제
- household appliance: 가전제품
- zinc oxide: 아연산화물
- surface coating : 표면처리
- outdoor furniture varnish : 실외가구 광택제
- cerium oxide : 산화세륨
- fuel catalyst : 연료 촉매

18 Car(Automobile)

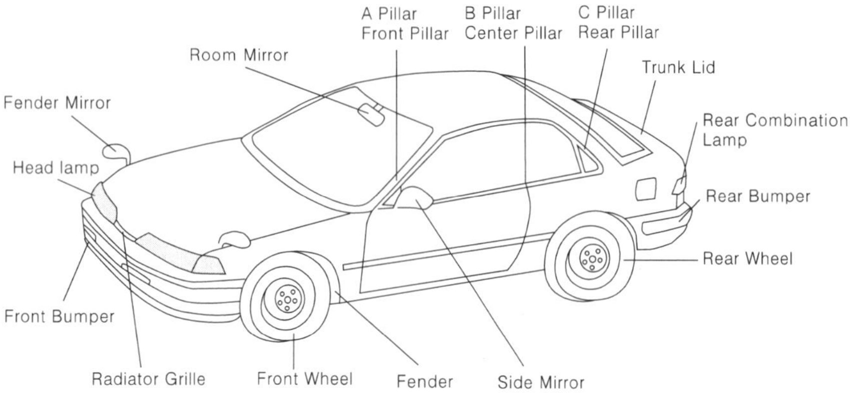

[Figure 18－1] Body

Water Cooling

Water cooling is a method of heat removal from components. As opposed to air cooling, water is used as the heat transmitter. Water cooling is commonly used for cooling internal combustion engines in automobiles and electrical generators. Other uses include cooling the lubricant oil of pumps; for cooling purposes in heat exchangers; cooling products from tanks or columns, and recently, cooling of various major components inside top-end personal computers. The main mechanism for water cooling is convective heat transfer.

Note

- cooling : 냉각
- removal : 제거
- oppose to ~ : ~에 반대하여
- transmitter : 전달장치
- internal combustion engine : 내연기관
- automobile : 자동차
- electrical generator : 전기발전기
- lubricant oil : 윤활유
- heat transfer : 열전달
- heat exchanger : 열교환기
- various : 다양한
- convective : 대류의

Coolant

A coolant, or heat transfer fluid, is a fluid which flows through a device in order to prevent its overheating, transferring the heat produced by the device to other devices that utilize or dissipate it. An ideal coolant has high thermal capacity, low viscosity, is low-cost, and is chemically inert, neither causing nor promoting corrosion of the cooling system. Some applications also require the coolant to be an electrical insulator. The coolant can either keep its phase and stay liquid or gaseous, or can undergo a phase change, with the latent heat adding to the cooling efficiency. The latter, when used to achieve low temperatures, is more commonly known as refrigerant.

Note

- coolant : 냉각제, 윤활제
- device : 장치
- dissipate : 열 따위를 방산하다
- viscosity : 점성
- inert : 비활성의
- corrosion : 부식
- insulator : 절연물, 절연체, 단열
- phase change : 상변화
- latent : 잠재적인
- efficiency : 효율
- refrigerant : 냉각제

Needle Valve

A needle valve is a type of valve having a small orifice and a threaded, needle-like plunger. It allows precise regulation of flow, although it is generally used for, and is capable of, only relatively small flow rates.

Note

- needle : 바늘, 나침
- plunger : 피스톤의 플런저, (긴 자루달린)흡인용 고무 컵
- precise : 정밀한
- regulation : 조절
- flow rate : 유량
- be capable of ~ : ~의 능력이 있는

Brake

A brake is a device for slowing or stopping the motion of a machine or vehicle, or alternatively a device to restrain it from starting to move again. The kinetic energy lost by the moving part is usually translated to heat by friction. Alternatively, in regenerative braking, much of the energy is recovered and stored in a flywheel, capacitor or turned into alternating current by an alternator, then rectified and stored in a battery for later use.

Note

- vehicle : 운송수단
- restrain : 억제하다
- kinetic energy : 운동 에너지
- be translated to ~ : ~으로 변환되다
- friction : 마찰
- alternator : 교류전원, 교류기
- capacitor : 콘덴서, 축전기
- alternating current : 교류
- rectify : 정류하다

Drum Brake

A drum brake is a brake in which the friction is caused by a set of shoes or pads that press against the inner surface of a rotating drum. The drum is connected to a rotating wheel.

Note

- friction : 마찰
- be connected to ~ : ~에 연결되어 있다

Brake Lining

Brake linings are the consumable surfaces in braking systems, especially those used in vehicles. Brake linings are composed of a relatively soft but tough and heat-resistant material with a high coefficient of dynamic friction (and ideally an identical coefficient of static friction) typically mounted to a solid metal backing using high-temperature adhesives or rivets.

Note

- lining : 라이닝(베어링, 실린더 따위의 안쪽에 댄 것)
- consumable : 소모품의
- tough : 단단한
- heat-resistant : 계수
- dynamic friction : 동마찰
- adhesive : 점착성의
- be composed of ~ : ~으로 구성되어 있다

Radiators

Radiators and convectors are types of heat exchangers designed to transfer thermal energy from one medium to another for the purpose of cooling and heating. The majority of radiators are constructed to function in automobiles, buildings, and electronics.

Note

- radiator : 라디에이터, 방열기, 냉각장치
- convector : 대류식방열기
- heat exchanger : 열교환기
- the majority of ~ : ~의 대부분은
- function : 기능

Transmission

Using the principle of mechanical advantage, transmissions provide a speed-torque conversion (commonly known as "gear reduction" or "speed reduction") from a higher speed motor to a slower but more forceful output.

Note

- transmission : 트렌스미션, 동력 전달 장치
- conversion : 변환
- reduction : 감소, 절감
- output : 출력

Shock Absorber

A shock absorber in common parlance (or damper in technical use) is a mechanical device designed to smooth out or dampen shock impulse, and dissipate kinetic energy. In a vehicle, it reduces the effect of traveling over rough ground, leading to improved ride quality. Without shock absorbers, the vehicle would have a bouncing ride, as energy is stored in the spring and then released to the vehicle, possibly exceeding the allowed range of suspension movement.

Note

- shock absorber : 완충기
- in common parlance : 일반적인 말로는
- impulse : 충격
- dissipate : 없어지다
- kinetic energy : 운동 에너지
- rough : 거친

- improve : 향상시키다
- bounce : 튀어오르다
- release : 해제하다
- exceeding : 초과하는

Front Engine Front Drive

In automotive design, a FF, or Front-engine, Front-wheel drive layout places both the engine and driven wheels at the front of the vehicle. This layout is typically chosen for its compact packaging, allowing the rest of the vehicle to be designed more flexibly. In contrast with the FR layout, the FF layout eliminates the need for a central tunnel or a higher chassis clearance to accommodate a driveshaft providing power to the rear wheels.

Note

- front engine front drive(FF) : 전륜 구동
- layout : 배치, 설계
- flexibly : 유연하게
- in contrast with ~ : ~과 대조적으로
- front engine rear drive(FR) : 후륜 구동
- eliminate : 제거하다
- accommodate : 수용하다

Antilock Brake System(ABS)

An anti-lock braking system (ABS) (translated from German, Anti-Blockier System) is a safety system on motor vehicles which prevents the wheels from locking while braking. A rotating road wheel allows the driver to maintain steering control under heavy braking, by preventing a locked wheel or skid, and allowing the wheel to continue to forward roll and create lateral control, as directed by driver steering inputs. Disadvantages of the system include increased braking distances under some limited circumstances (ice, snow, gravel, "soft" surfaces), and the creation of a "false sense of security" among drivers who do not understand the operation, and limitations of ABS.

Note

- antilock brake system : 앤티록식 브레이크 장치
- prevent : ~을 예방하다, 방지하다
- maintain : 유지하다
- steering : 조종
- skid : 미끄러짐
- lateral : 측면의
- disadvantage : 단점
- gravel : 자갈

Convertible

Convertible is an automobile body style with a flexibly operating roof that can convert between open-air or enclosed modes. Roof designs are highly variable from the folding textile roof, known variously as the top, soft top, ragtop or hood- to the folding multi-sectional rigid roof known as a retractable hardtop, coupé convertible or coupé cabriolet.

Note

- convertible : (자동차가)접는 포장이 달린
- enclosed : 둘러싸다
- fold : 접다
- textile : 직물의
- variously : 여러 가지의
- retractable : 신축자재의
- ragtop : 포장 지붕식 자동차
- coupé : 2인승 4륜 유개마차

Concept Car

A concept car or show car is a car prototype made to showcase a concept, new styling, technology and more. They are often shown at motor shows to gauge customer reaction to new and radical designs which may or may not have a chance of being produced.

Note

- concept car : 컨셉카
- prototype : 원형, 본
- motor show : 자동차 전시회
- radical : 근본적인, 급진적인, 급격한

Common Rail Direct Injection(CRDI)

Common rail direct fuel injection is a modern variant of direct fuel injection system for diesel engines. It features a high-pressure (over 1,000übar/15,000üpsi) fuel rail feeding individual solenoid valves, as opposed to low-pressure fuel pump feeding unit injectors (Pumpe Düse or pump nozzles), or high-pressure fuel line to mechanical valves controlled by cams on the camshaft. Third-generation common rail diesels now feature piezoelectric injectors for increased precision, with fuel pressures up to 1,800bars (26,000psi).

Note

- injection : 분사
- variant : 변형
- feature : 특징으로 삼다
- piezoelectric injector : 압전 분사기
- compliance : 승낙, 준수

Turbocharged Direct Injection(TDI)

Turbocharged Direct Injection (TDI) is the technology used to describe, and name modern turbocharged diesel engines produced by Volkswagen Group (Volkswagen AG) (VWAG), and widely used in all marques of passenger cars and light commercial vehicles produced by the company (particularly those sold in Europe).

Note

- turbocharger : 배기 터빈 과급기
- marque : 차종
- passenger car : 승용차
- commercial : 상업적인

19 Robotics

Robotics

Robotics is the science and technology of robots, their design, manufacture, and application. Robotics requires a working knowledge of electronics, mechanics and software, and is usually accompanied by a large working knowledge of many subjects.

The structure of a robot is usually mostly mechanical and can be called a kinematic chain (its functionality being similar to the skeleton of the human body). The chain is formed of links (its bones), actuators (its muscles) and joints which can allow one or more degrees of freedom.

Robots used as manipulators have an end effector mounted on the last link. This end effector can be anything from a welding device to a mechanical hand used to manipulate the environment. ISO 10248 defines a robotic application on the industrial field.

Note

- design : 설계
- manufacture : 제작
- application : 응용
- accompany : 동반하다
- kinematic chain : 연쇄
- skeleton : 골격
- link : 링크, 연쇄
- degrees of freedom : 자유도
- manipulator : 조종기
- end effector : 말단 조작기

Origin of robot

According to the Oxford English Dictionary, the word robotics was first used in print by Isaac Asimov, in his science fiction short story "Liar!", published in May 1941 in Astounding Science Fiction. Robotics is based on the word "robot" coined by science fiction author Karel Capek in his 1920 theater play R.U.R. (Rossum's Universal Robots). The word robot comes from the word robota meaning "self labor", and, figuratively, "drudgery" or "hard work" in Czech (and many other Slavic languages). Asimov was unaware that he was coining the term for a new field-as the design of electrical devices is called electronics, so the design of robots could be appropriately called robotics. Robots are used in industrial, military, exploration, home making, and academic and research applications.

Note

- according to ~ : ~에 의하면
- Isaac Asimov : 러시아의 소설가
- astound : 몹시 놀라게 하다
- Astounding Science Fiction : 미국의 SF잡지
- coin : (새로운 표현을) 만들어내다
- Karel Capek : 체코의 극작가
- figuratively : 비유적으로
- drudgery : 단조로운 고역
- unaware : 알지 못하는
- appropriately : 적절하게

Configuration of industrial robots

The mechanical construction of industrial robots is based on these basic movements, i.e. robots have one or more translational axes and/or rotational axes. The more axes of motion a robot has, the more mobile it is; it is said that a robot has as many degrees of freedom as it has axes of motion.

The most widely known industrial robots are composed of 3 main axes of motion(3 degrees of freedom). The movement of the main axis is also designated as the macro movement.

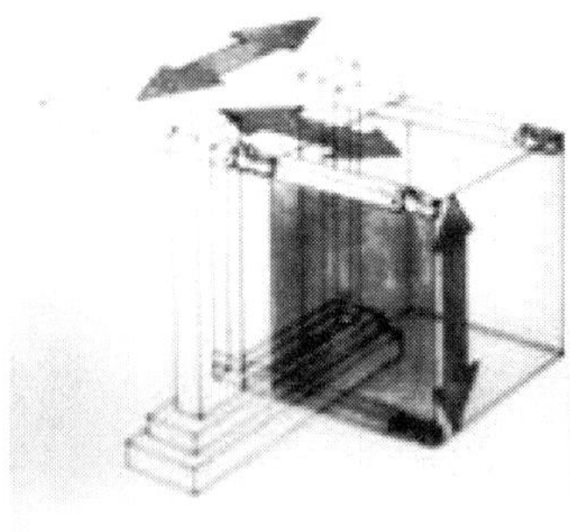

[Figure 19-1] Robot with cartesian working area

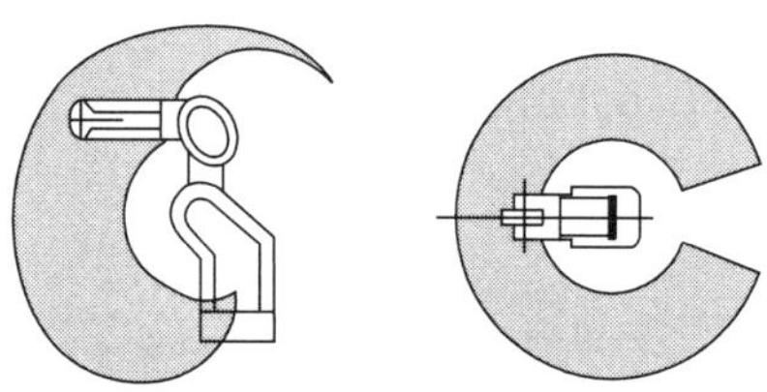

[Figure 19-2] Robot with spherical working area

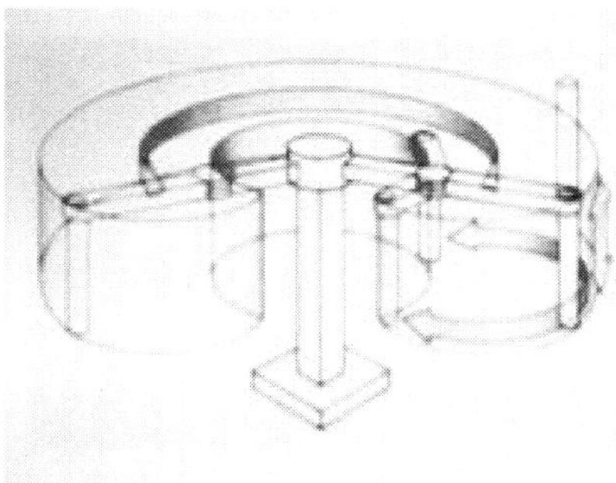

[Figure 19-3] Robot with cylindrical working area

In order that the industrial robot can, for example, grip a workpiece, it must have a manipulator at the end of the chain of movement (kinematic chain). This must also be able to move in order that it can originate itself towards objects. The movement of the manipulator is also designated as the micro movement. The manipulator also usually has several degrees of freedom.

Note

- translational : 병진운동의
- axes : axis의 복수, 축
- rotational axes : 회전 축
- degrees of freedom : 자유도
- compose : 구성하다
- designate : 가리키다
- macro movement : 주요 운동
- cartesian : 직각좌표계의
- spherical : 구형의
- cylindrical : 원통형의
- workpiece : 제조 공정에 있는 제품이나 소재
- manipulator : 조종기
- kinematic chain : 연쇄
- originate : 시작하다
- degrees of freedom : 자유도

Actuation

A robot leg, powered by Air Muscles. The actuators are the 'muscles' of a robot; the parts which convert stored energy into movement. By far the most popular actuators are electric motors, but there are many others, some of which are powered by electricity, while others use chemicals, or compressed air.

By far the vast majority of robots use electric motors, of which there are several kinds. DC motors, which are familiar to many people, spin rapidly when an electric current is passed through them. They will spin backwards if the current is made to flow in the other direction.

The air muscle is a simple yet powerful device for providing a pulling force. When inflated with compressed air, it contracts by up to 40% of its original length. The key to its behavior is the braiding visible around the outside, which forces the muscle to be either long and thin, or short and fat. Since it behaves in a very similar way to a biological muscle, it can be used to construct robots with a similar muscle/skeleton system to an animal. For example, the Shadow robot hand uses 40 air muscles to power its 24 joints.

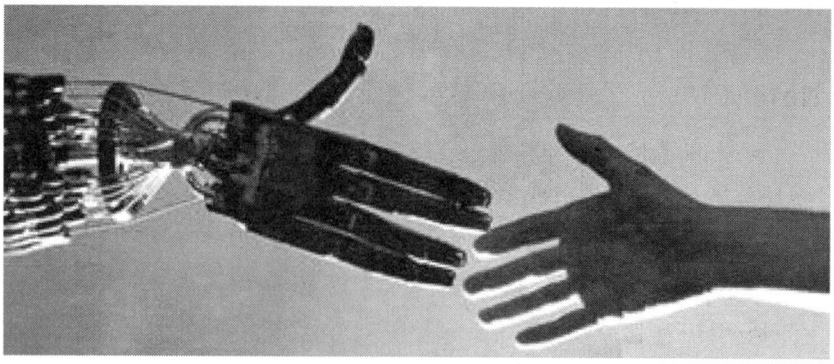

[Figure 19-4] Shadow robot hand

Note

- air muscle : 공기압을 이용한 근육장치
- actuator : 액추에이터, 구동장치
- by far : 매우
- majority : 대부분
- backward : 역방향으로
- contract : 수축시키다
- braid : 짜다
- convert : 전환하다
- compressed air : 압축공기
- spin : 회전시키다
- inflate : 팽창시키다
- behavior : 작용
- biological : 생물학적

Manipulation

Robots which must work in the real world require some way to manipulate objects; pick up, modify, destroy or otherwise have an effect. Thus the 'hands' of a robot are often referred to as end effectors, while the arm is referred to as a manipulator. Most robot arms have replaceable effectors, each allowing them to perform some small range of tasks. Some have a fixed manipulator which cannot be replaced, while a few have one very general purpose manipulator, for example a humanoid hand.

A common effector is the gripper. In its simplest manifestation it consists of just two fingers which can open and close to pick up and let go of a range of small objects.

Some advanced robots are beginning to use fully humanoid hands, like the Shadow Hand. These highly dexterous manipulators, with as many as 20 degrees of freedom and hundreds of tactile sensors can be difficult to control.

Note

- manipulate : 조종하다
- end effector : 말단 조작기
- task : 작업
- gripper : 그리퍼, 잡는 기구
- dexterous : 솜씨 좋은
- destroy : 분쇄하다
- replaceable : 바꾸어 놓을 수 있는
- humanoid : 인간형
- manifestation : 표현
- tactile sensor : 촉각센서

Control

A robot-manipulated marionette, with complex control systems. The mechanical structure of a robot must be controlled to perform tasks. The control of a robot involves three distinct phases - perception, processing and action (robotic paradigms). Sensors give information about the environment or the robot itself (e.g. the position of its joints or its end effector). This information is then processed to calculate the appropriate signals to the actuators (motors) which move the mechanical structure.

The processing phase can range in complexity. At a reactive level, it may translate raw sensor information directly into actuator commands. Sensor fusion may first be used to estimate parameters of interest (e.g. the position of the robot's gripper) from noisy sensor data. An immediate task (such as moving the gripper in a certain direction) is inferred from these estimates. Techniques from control theory convert the task into commands that drive the actuators.

Note

- manipulate : 조종하다
- task : 작업
- perception : 인식, 인지
- end effector : 말단 조작기
- processing phase : 신호처리단계
- sensor fusion : 센서 통합 시스템
- marionette : 꼭두각시
- distinct : 별개의
- paradigm : 보기
- appropriate : 적절한
- raw : 가공하지 않은
- infer : 추론하다

Sensors

A sensor is a device that measures some attribution of the world. Being one of the three primitives of robotics (besides processing and action), sensing plays an important role in robotic paradigms.

1. Proprioceptive Sensors

Proprioceptive sensors sense the position, the orientation and the speed of the humanoid's body and joints.

In human beings inner ears are used to maintain balance and orientation. Humanoid robots use accelerometers to measure the acceleration, from which velocity can be calculated by integration; tilt sensors to measure inclination; force sensors placed in robot's hands and feet to measure contact force with environment; position sensors, that indicate the actual position of the robot (from which the velocity can be calculated by derivation) or even speed sensors.

Note

- attribution : 속성
- besides : 그밖에
- paradigm : 보기
- orientation : 방향
- integration : 적분
- inclination : 기울기
- primitive : 근본의
- processing : 신호처리
- proprioceptive sensors : 내계 센서
- humanoid robot : 인간형 로봇
- tilt sensor : 기울기 센서
- derivation : 미분

2. Exteroceptive Sensors

Exteroceptive sensors give the robot information about the surrounding environment allowing the robot to interact with the world.

Proximity sensors are used to measure the relative distance between the sensor and objects in the environment. Tactile sensors also provide information about forces and torques transferred between the robot and other objects.

Vision refers to processing data from any modality which uses the electromagnetic spectrum to produce an image. In humanoid robots it is used to recognize objects and determine their properties. Vision sensors work most similarly to the eyes of human beings. Most humanoid robots use CCD cameras as vision sensors.

Sound sensors allow humanoid robots to hear speech and environmental sounds, and perform as the ears of the human being. Microphones are usually used for this task.

Note

- exteroceptive sensors : 외계 센서
- surrounding environment : 주위 환경
- interact : 상호작용하다
- proximity sensor : 근접 센서
- tactile sensor : 촉각 센서
- modality : 양상
- humanoid robot : 인간형 로봇
- recognize : 인식하다
- CCD : charge-coupled device
- vision sensor : 시각센서
- sound sensor : 음파센서

Dynamics and kinematics

The study of motion can be divided into kinematics and dynamics. Direct kinematics refers to the calculation of end effector position, orientation, velocity and acceleration when the corresponding joint values are known. Inverse kinematics refers to the opposite case in which required joint values are calculated for given end effector values, as done in path planning. Some special aspects of kinematics include handling of redundancy(different possibilities of performing the same movement), collision avoidance and singularity avoidance. Once all relevant positions, velocities and accelerations have been calculated using kinematics, methods from the field of dynamics are used to study the effect of forces upon these movements.

Direct dynamics refers to the calculation of accelerations in the robot once the applied forces are known. Inverse dynamics refers to the calculation of the actuator forces necessary to create prescribed end effector acceleration. This information can be used to improve the control algorithms of a robot.

Note

- kinematics : 운동학
- end effector : 말단 조작기
- corresponding : 대응하는
- path planning : 경로계획
- redundancy : 중복성
- avoidance : 회피
- relevant : 관련된
- prescribe : 지시하다
- dynamics : 동역학
- orientation : 좌표
- inverse kinematics : 역기구학
- aspect : 견해
- collision : 충돌
- singularity : 특이성
- inverse dynamics : 역동역학

Humanoid robot

HUBO, an example of a humanoid robot. A humanoid robot is a robot with its overall appearance based on that of the human body. In general humanoid robots have a torso with a head, two arms and two legs, although some forms of humanoid robots may model only part of the body, for example, from the waist up. Some humanoid robots may also have a 'face', with 'eyes' and 'mouth'. Androids are humanoid robots built to resemble a male human, and Gynoids are humanoid robots built to resemble a human female.

[Figure 19-5] HUBO

Note

- humanoid robot : 인간형 로봇
- HUBO : 휴머노이드(Humanoid)와 로봇(Robot)의 합성어, 2004년 한국과학기술원(KAIST)에서 만든 인간형 로봇
- appearance : 외관
- torso : 인간의 몸통
- resemble : ~을 닮다

공업영어

2008년 8월 25일 제1판 1쇄 발행
2018년 9월 15일 제1판 5쇄 발행

저　　자 ◎ 김지환 · 김 중 · 김도석 · 이창훈

발 행 인 ◎ 조승식

발 행 처 ◎ (주)도서출판 북스힐
서울시 강북구 한천로 153길 17

등　　록 ◎ 제22-457호(1998년 7월 28일)

(02) 994-0071(代)

(02) 994-0073

bookshill@bookshill.com
www.bookshill.com

값 12,000원

ISBN 978-89-5526-517-0